Tank Manual
Normandy 1944

AMBERLEY

First published 2018

Amberley Publishing
The Hill, Merrywalks
Stroud, Gloucestershire, GL5 4EP

www.amberley-books.com

Copyright © Amberley Archive, 2018

The right of Amberley Archive to be identified as the Author
of this work has been asserted in accordance with the
Copyrights, Designs and Patents Act 1988.

All rights reserved. No part of this book may be reprinted
or reproduced or utilised in any form or by any electronic,
mechanical or other means, now known or hereafter invented,
including photocopying and recording, or in any information
storage or retrieval system, without the permission in writing
from the Publishers.

British Library Cataloguing in Publication Data.
A catalogue record for this book is available from the British Library.

ISBN 978-1-4456-7835-1

Typesetting and Origination by Amberley Publishing.
Printed in Great Britain.

CONTENTS

Introduction	4
Driver's Handbook for the Churchill Infantry Tank	7
Field Manual for the M5 Stuart Light Tank	92
Field Manual for the M4 Sherman Medium Tank	177
Cromwell Tank Service Instruction Book	250

INTRODUCTION

During the early years of the Second World War, Britain was under extreme pressure to arm itself. There was a lack of military hardware with which to meet the Nazi threat, and this was reflected by the armed forces' shortage of tanks.

In the circumstances, there was no time to produce a next-generation tank from scratch. Instead, the A22 'Churchill' was developed rapidly, based around an earlier model. The Churchill therefore had more shortcomings than a newer tank would have. Notably, it lacked speed, because it was designed to support the infantry and move alongside them. But it was well-armoured and had the advantage of being quick to production. The Churchill might not have been a match for the best German machinery, but the pragmatic approach was essential: it was better to have tanks ready for action than to be crucially short of them. As with so much in wartime, it was dictated by necessity.

Due to the lack of time for testing, ongoing improvements were made when resources allowed and as problems came to the notice of tank crew and engineers. The benefit of consistently improving an established design was reliability, which the Churchill had achieved by the middle of the war. Some important improvements, such as increased firepower, were delayed because factories were too busy producing the existing model. But by 1944, the Churchill Mark VII combined a wider chassis with a 75 mm gun.

Churchill tanks were also given specialist modifications, resulting in the 'Crocodile' variant armed with a flame-thrower, the Assault Vehicle Royal Engineers (AVRE) armed with mortar bombs, and others converted to ploughs and rollers, those used to clear mines, to place explosives, and to carry mobile bridges for spanning watercourses or tank traps. So, despite its limitations, the Churchill was ingeniously adapted and important to the British Army in active service. Later, in time for D-Day, Sherman tanks would be adapted in similar ways.

The less favoured Cromwell tank was, like the Churchill, being fitted with a 75 mm gun by 1944, and weaker components were being eliminated, to improve its performance in crucial areas such as shooting, strength against mines, and water resistance.

However, the British needed more than the Churchill and an improved Cromwell. The evidence pointed to ever more effective enemy tanks that might tip the balance on the battlefield in future. For example, the German Tiger tank, which lacked reliability and ease of manufacture, nonetheless had formidable fire-power and 4½ inch armour plating, while the gun of a German Panther tank was better at penetrating armour than those of either a Cromwell or Sherman.

Yet the mass production of the Sherman enabled Britain to take around 7,000 of the M4A4 variant by late 1943 so, combined with its Churchills and Cromwells, Britain played its part on amassing an overwhelming Allied superiority in tank numbers in the build-up to the invasion of Normandy. Importantly, each of these three tanks was a known quantity, available with many adaptations for different battlefield requirements, and with technical problems relatively well ironed out. As a result, the Allies might not have possessed the best tanks, but they had an undeniable advantage in practice.

By June 1944, the British and Americans had assembled history's biggest and heaviest invasion force, but great risks still remained. The Allied forces now faced a unique challenge of transporting and landing their tanks and men on Normandy's sandy beaches. They would need more than superior numbers to break through the Atlantic Wall. As tank crew prepared for the invasion, they relied on the manuals and instructions reproduced in this book, to maintain and operate their tanks during D-Day and the many subsequent days of battle. These guides to the main tanks deployed in 1944 provide a fascinating insight into the vehicles' engineering and the practical realities of taking them into battle.

A British Sherman Firefly tank, modified to carry a 17-pounder gun, seen in 1944. (US Army)

DRIVER'S HANDBOOK
for the
"CHURCHILL"
INFANTRY
TANK

FOREWORD

This handbook provides, in brief form and in handy pocket size, a "driver's guide" to the Churchill Infantry Tank. It is not intended to replace the Official Instruction Book (which is, of course, compiled in greater detail), but is simply a resume of all the vital information required by the driver in a more convenient format for "on the spot" reference. It describes, as simply as possible, the controls, method of driving and general handling of the vehicle (pages 3 to 26). It lists dimensions, petrol, oil and water capacities and other data required by the crew (pages 59 and 60). It includes instructions for general lubrication and straightforward routine maintenance (pages 61 to 80). And it contains a short but important list of the points which need a little *extra* care and attention if trouble is to be avoided.

A complete list of the items of stowage for all models (Churchill I, II, III and IV), fully illustrated, appear on pages 27 to 58. The stowage illustrations also provide good general views of the various models.

The maintenance section—pages 61 to 80—is divided into periods ("Daily," "Every 250 Miles," and so on) with a special section showing the operations which need to be done more frequently in dusty conditions. Lubrication and topping-up jobs are printed in ordinary type, and inspection and adjustment items are in capital letters. Most of the routine jobs necessary are illustrated, and clear references are quoted for those which entail reference to the Instruction Book proper.

In short, the Driver's Handbook contains the basic information necessary to drive and maintain the vehicle. When fuller or more detailed information is required, reference should always be made to the Instruction Book.

CONTROLS

MASTER SWITCH

Located in the battery recess. Provided to cut off all battery current from the vehicle. A green warning light glows on the instrument panel when it is on.

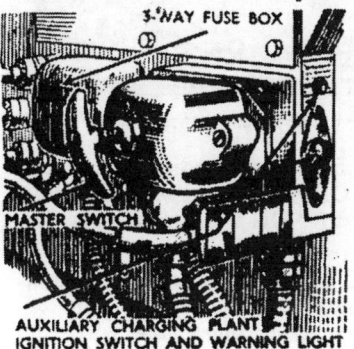

Fig. 1. Battery master switch.

IGNITION SWITCH

Mounted on the instrument panel. A red warning light adjacent to it glows all the time it is switched on. The master switch (see paragraph above) must be switched on first.

STARTER BUTTON

Located in the centre of the instrument panel, and operated by pushing. See detailed instructions for starting engine on page 16.

AMMETER

Situated on the left of the instrument panel. The charging rate shown when the engine is running will vary considerably —high when the engine is started; gradually lower as the battery is recharged. Does not register discharge.

PETROL GAUGES (2)

Near the bottom right-hand corner of the instrument panel —one for right-hand tanks; one for left-hand tanks. The gauges are electrically operated and record the petrol level only when the push button just above them is pressed.

PETROL CONTROLS (3)

Main Tanks Control. Fitted on the left-hand "wall" of the driving compartment, near the roof. To shut off supply, move lever to extreme right. To draw on left-hand tanks move lever to "L H." To draw on right-hand tanks move lever to "R.H."

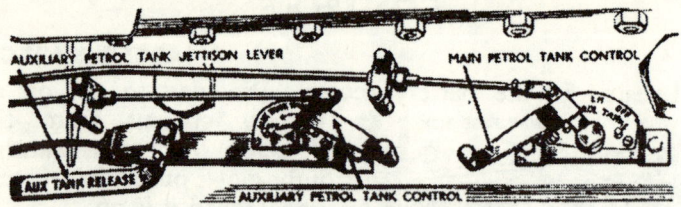

Fig. 2. Main and auxiliary petrol tank controls and jettison release lever.

Auxiliary Tank Control. Immediately adjacent to main tanks control. Keep lever in " OFF " position when drawing from main tanks. Move lever to " ON " to draw from the auxiliary tank.

Jettison Control. Situated next to the auxiliary tank control. Operates release mechanism, *and should only be moved when it is desired to jettison the auxiliary tank.*

PRIMING CONTROL

On the fighting compartment rear bulkhead—top right-hand corner. Provided so that the carburettors can be hand-primed before starting the engine. Pull out (to full extent) and push in several times before starting from cold.

" KI-GASS " PUMP CONTROL

Operating knob is mounted on the fighting compartment rear bulkhead. Three or four strokes will prime the engine

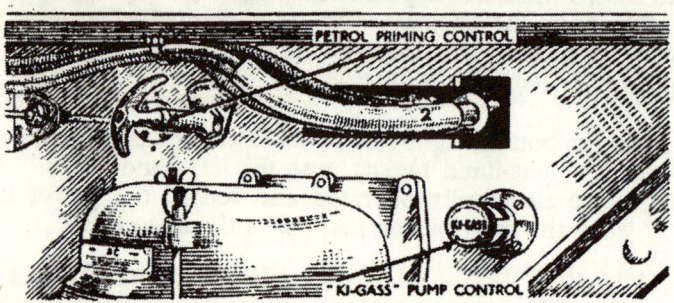

Fig. 3. Priming control and " Ki-gass " pump operating plunger.

4

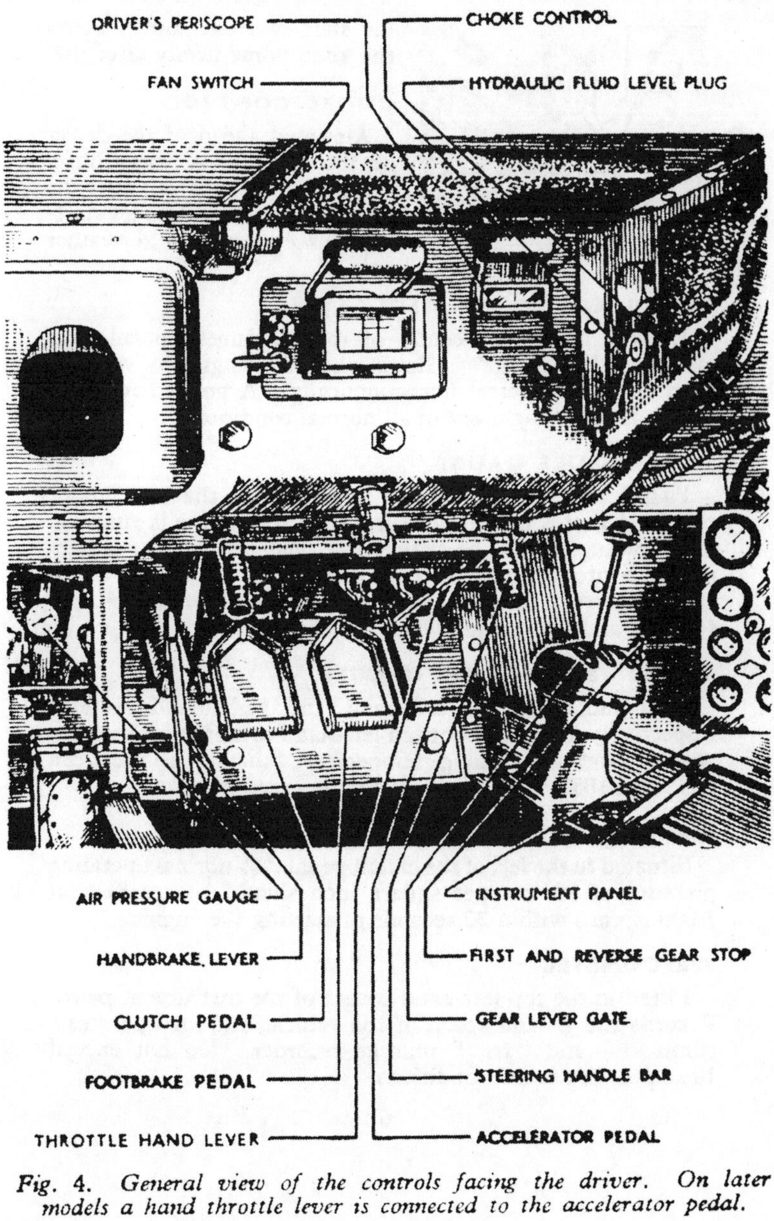

Fig. 4. General view of the controls facing the driver. On later models a hand throttle lever is connected to the accelerator pedal.

Fig. 5. Carburettor choke control.

for starting from cold. Screw the knob home firmly after use.

CHOKE CONTROL

Mounted ahead of the driver on the side wall of the vehicle. Operates the choke valves of the four carburettors. Use for starting in exceptionally cold weather only.

PETROL ECONOMY LIGHT

Located at the top-centre of the instrument panel, and labelled "PETROL." Glows when the engine is working hard and using petrol uneconomically. A good driver will aim to keep the light *out* in all normal conditions.

OIL PRESSURE GAUGE

Placed to the right of the ammeter, on the instrument panel. Records the oil pressure when the engine is running. The reading should be between 40 and 80 lbs. per square inch. Switch off and report immediately if it falls below 15 lbs. per sq. inch at normal idling speed, or 40 lbs. per sq. inch at 2,000 r.p.m., when the engine is hot.

WATER TEMPERATURE GAUGES

Two indicators for left-hand and right-hand systems respectively, located at bottom left-hand corner of instrument panel. Normal running temperature should be between 165° and 185°. Report if reading is excessive.

AIR PRESSURE GAUGE

Situated to the left of the clutch pedal. A normal operating pressure of 80 lbs. per square inch should be reached (at 1,000 r.p.m.) within 30 seconds of starting the engine.

SPEEDOMETER

Fitted in the top left-hand corner of the instrument panel. Records the ground speed of the vehicle, and incorporates a cumulative and "trip" mileage recorder. Do not exceed 10 m.p.h. in normal conditions.

TACHOMETER

This is a revolution indicator, mounted immediately below the speedometer. It registers the engine speed in multiples of 100 revolutions per minute. Reading should not exceed 2,000 in normal driving conditions. Watch this point carefully when descending hills, and do not allow the engine speed to rise above the recommended maximum.

ACCELERATOR PEDAL

The right-hand of the three foot controls. Operates the throttles of the four carburettors through a hydraulic control to the engine compartment and thence by a system of interconnected rods and levers. A right-angled lever is connected to the accelerator pedal for hand throttle operation.

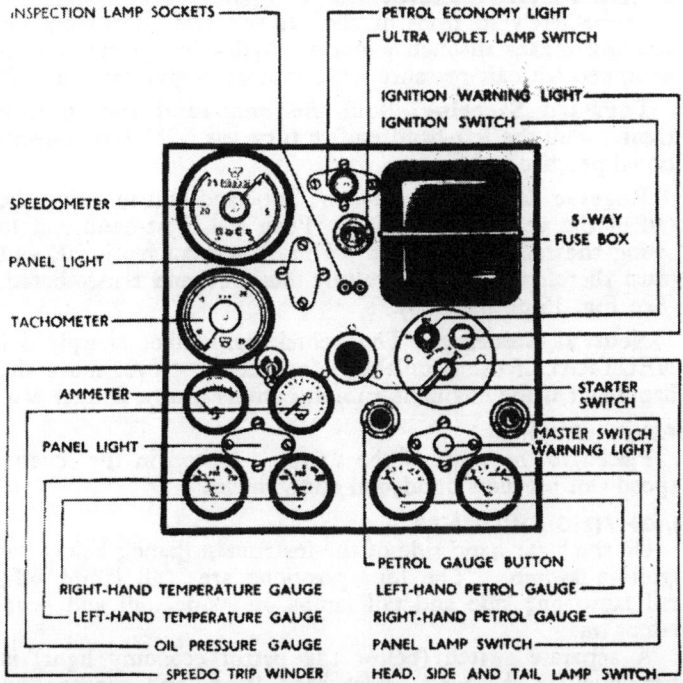

Fig. 6. . General view of the switches, gauges and other instruments on the instrument panel. A convoy lamp switch is fitted to later vehicles.

FOOTBRAKE PEDAL
The centre foot control. Operates the main (stopping) brakes through a conventional Lockheed hydraulic system.

CLUTCH PEDAL
The left-hand foot control. Operates the clutch hydraulically and is linked up with an air pressure servo motor to provide a light pedal action.

HANDBRAKE LEVER
Situated to the left of the three foot pedals. Used as a parking brake to apply the main (stopping) brakes. If the vehicle is to be parked on a gradient exceeding 1 in 10 engage first speed gear.

STEERING HANDLEBAR
Immediately in front of the driver's seat. Operates the steering brakes through a normal hydraulic system, and is equipped with air pressure servo motors to give easy control.

Forward Steering. Pull the right-hand end to turn right. Pull the left-hand end to turn left. This is conventional practice.

Reverse Steering. *Push* the left-hand end to swing the tail of the vehicle to the left. Push the right-hand end to swing the tail to the right. This is *not* conventional, and must therefore be very carefully mastered and remembered. (See Fig. 15 on page 23).

Neutral Steering. The vehicle will pivot sharply IN NEUTRAL if the handlebar is operated. Never move the handlebar if the engine is running unless a turn is intended.

GEAR LEVER
Placed to the right of the driver. A stop in the reverse speed slot prevents accidental engagement.

LIGHTING SWITCHES
On the right-hand side of the instrument panel, below the ignition switch. The four positions are: all lamps off; tail lamp on; side and tail lamps on; side, tail and head lamps on.

A separate switch (below the petrol economy light) is provided for the ultra violet head lamp, and another, on recently produced vehicles, for the convoy lamp.

The instrument panel lamps are controlled by a switch fitted below and to the right of the driving lights switch.

INTERIOR LIGHTS

Driving Compartment. Two lights are fixed to the roof, one behind the driver's seat and one behind the front gunner's seat. A switch is fitted to the base of each light.

Fighting Compartment. Three lights are fitted to the roof of the turret. Each has a switch incorporated in the base.

TELEPHONE POINTS

Driving Compartment. The driver's telephone connection is behind the driver's seat on the spot-light bin. The front gunner's connection is fixed to the roof, above and behind the front gunner's seat.

Fighting Compartment. The connection for the commander's and turret gunner's telephone is on the left-hand side of the turret, just below the cupola. The wireless operator's connection is on the right-hand side of the turret, just under the escape hatch.

VENTILATING FANS

Driving Compartment. On vehicles equipped with a Besa gun in the driving compartment, a ventilating fan is fitted immediately to the left of the driver's vision port. The switch is on the junction box near the fan.

Fighting Compartment. On vehicles fitted with a turret, a fan is mounted on the roof immediately above the Besa. The switch is on the junction box just above the turret traverse control.

DRIVER'S VISION PORT

The large door is opened by a lever just above the steering handlebar. The same lever operates the wicket door once the locking catch is released. Spring-loaded catches lock both doors in the open position, and a latch locks them when closed. All controls can be operated from inside the vehicle. To give protection to the driver when the front vision door is open, a detachable window clips into the opening.

PERISCOPES

A periscope is provided for the driver to use when the vision port is completely closed—see above. A similar periscope is provided for the front gunner.

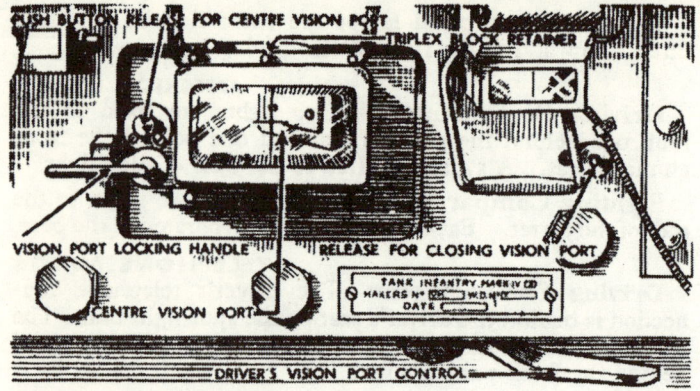

Fig. 7. Driver's vision port and periscope

ESCAPE HATCHES
Double-hinged doors provided in the roof above the driver and front gunner. They are fastened from the inside, but can be opened or fastened from the outside with a suitable key.

PANNIER DOORS
One in each pannier, opening into the driving compartment. These doors can be opened or fastened only from the inside. The locking device acts automatically when they are pulled to.

REVOLVER PORTS
One in each pannier door. The ports are hingeless, and are fitted with a rapid opening and closing control. Each has a locking bolt to secure it in the closed position. Similar revolver ports (two) are provided in the turret.

SMOKE MORTAR GENERATOR (on later vehicles)
The push-buttons for operating the smoke screen are on a panel mounted above the instrument panel.

MUD PLOUGHS (on later vehicles)
Fitted below the final drives, they prevent damage to the track guards by preventing mud packing in "heavy going" conditions. When not required, the ploughs can be raised from the tracks.

EMERGENCY CONTROLS

EXTRA STEERING CONTROL

A half handlebar connected to the main steering handlebar and located in front of the front gunner. This control enables the gunner to steer the vehicle in an emergency.

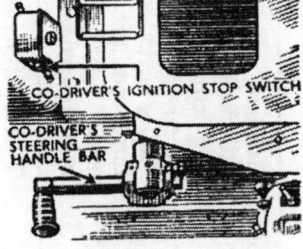

Fig. 8. *Emergency steering control and ignition cut-off switch facing co-driver's seat.*

IGNITION CUT-OFF SWITCH

On the left of the front gunner. Provided so that the gunner can switch off the engine if the necessity arises. Normally, this switch should not be touched. It is not fitted to some vehicles with a 3" howitzer in the driving compartment.

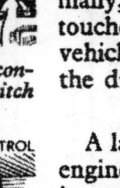

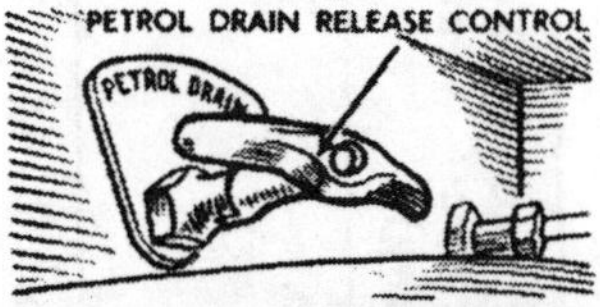

Fig. 9. *Pull-out handle controlling petrol release trap.*

PETROL RELEASE

A large trap in the floor of the engine compartment, operated by a pull-out handle on the rear bulkhead (near the roof) of the fighting compartment. Through this trap, escaped petrol can be quickly jettisoned.

EMERGENCY DISPOSAL HATCH

In the floor, behind the front gunner's seat. Fastened by a long-handled screw, it can be swung to one side to jettison anything not required.

Fig. 10. *CO_2 operating lever and safety spring.*

FIRE EXTINGUISHER OUTFIT

2 CO_2 bottles on the rear centre of the fighting compartment, connected by pipes to nozzles in the engine and gearbox compartments. To operate, pull up the lever on top of the bottle.

IT IS IMPERATIVE THAT THE ENGINE BE SWITCHED OFF BEFORE OPERATING THE FIRE EXTINGUISHERS.

11

The lever is retained by a safety catch which should always be removed before starting the engine.

Hand extinguishers are also provided—two on the gearbox inspection doors, one on the turret gunner's leg shield and one on the side of the miscellaneous stowage bin (front left-hand side).

FUSES

Several fuses are provided in the electrical system to protect the wiring and components.

If one or more of the electrical units fails to function, check the appropriate fuse, and if it is "blown" make sure that there is no short circuit in the system before fitting a new one.

The list on this and the facing page shows the size, location and type of each fuse, the circuit it protects, and where to look for the spares.

LOCATION AND DETAILS OF FUSES

Location	Size	Type	Location of Spares	Circuits Protected
Battery Recess (on voltage regulator).	120 amps.	Strip tin	In cover	Main regulator to dynamo.
Battery Recess (on battery panel in 3-way fuse box) (*NOTE: On earlier models the spare fuse wire in the Battery Recess is all of one gauge. Two thicknesses of this wire should be used when renewing the centre fuse*).	Top: 20 amps.	Bridge and loose wire (34 swg)	Wound round bridge	Auxiliary charging and subsidiaries (coil, warning light and petrol pump).
	Centre: 60 amps.	Do. (27 swg)	Do.	Radio, turret interior lights, turret spotlight, forward interior light and forward communications.
	Bottom: 60 amps.	Do. (27 swg)	Do.	Turret traverse excitation and turret ventilating fan.

Instrument Panel (in 5-way fuse box).	A. 20 amps.	Bridge and loose wire (34 swg)	Wound round bridge	Ultra violet lamp, forward vent fan and infantry gong.
	B. 20 amps.	Do.	Do.	Inspection lamp socket, petrol gauges, panel lights and convoy lamp.
	C. 20 amps.	Do.	Do.	Tail lamp.
	D. 20 amps.	Do.	Do.	Side lamps.
	E. 20 amps.	Do.	Do.	Plain headlamp.
Turret (in turret spotlight switch box)	6 amps.	Glass cartridge	In switch box	Turret interior lights and turret spotlight.
Driver's Compartment (in telephone and lights junction box).	6 amps.	Glass cartridge	In junction box	Forward interior lights.
Wireless Telegraph Set and Power Unit (2 fuses incorporated in radio).	Special	Glass cartridge	In wireless kit	Radio.

HOW TO DRIVE

The first part of this chapter (pages 14 to 19) explains (a) what should be done *before* starting the engine, (b) how the engine should be started, and (c) the checks and inspections that should be carried out as soon as the engine is running and before the vehicle is driven off.

In each case, the instructions are detailed in numerical sequence on one page and illustrated on the facing page.

BEFORE STARTING

The following operations should always be carried out before starting the engine. They are clearly illustrated in Fig. 11 on the facing page.

1. **Check the Engine Oil Level.** The dipstick is under the screw-type filler cap on top of the de-aerator (on the right-hand side of the engine towards the rear). On early models the dipstick is located in a tube alongside the de-aerator. Check with the vehicle on level ground.

2. **Check the Petrol Tank Levels.** Two filler caps in the gearbox compartment, one each side at the front. The tanks are full when the level is $4\frac{1}{2}$ in. below the filler caps. Do not replenish above this point.

3. **Check Water Levels.** Two filler caps in the engine compartment, one each side at the front. Securely fasten the caps after checking, as the system operates under pressure.

4. **Release CO_2 Safety Catches.** There is a safety catch on each of the CO_2 fire extinguisher bottles. The bottles are located on the fighting compartment rear bulkhead.

Finally, make certain that the gear lever is in neutral, the steering bar straight, the handbrake on, and THAT NOBODY IS UNDER OR NEAR THE VEHICLE.

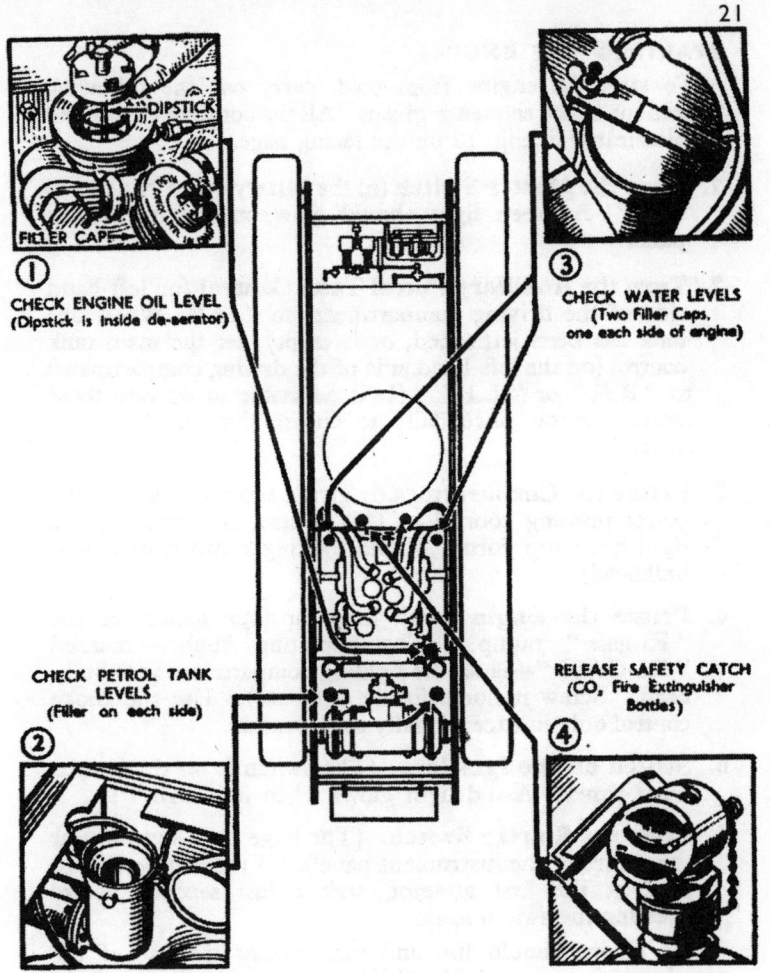

Fig. 11. Check these items before attempting to start the engine. They are explained (numerically) on the facing page.

15

STARTING THE ENGINE

To start the engine from cold, carry out the following operations in the sequence given. All the controls mentioned are illustrated in Fig. 12 on the facing page.

1. **Turn the Master Switch** (in the battery compartment) to "on." A green light should glow on the instrument panel.

2. **Turn the Auxiliary Petrol Tank Control** (on left-hand side of the driving compartment) to "on." When this tank has been jettisoned, or is empty, set the main tank control (on the left-hand side of the driving compartment) to "R.H." or "L.H." (It is advisable to operate these petrol controls once daily to ensure that they function correctly).

3. **Prime the Carburettors** by giving several strokes to the petrol priming control. (This control is located in the right-hand top corner of the fighting compartment rear bulkhead).

4. **Prime the Engine** with three or four strokes of the "Ki-gass" pump. (The operating knob — marked "KI-GASS"—is on the fighting compartment rear bulkhead. Screw it home firmly after use). Use the choke control only in exceptionally cold weather.

5. **Switch on the Ignition.** (The switch is on the instrument panel. A red light glows when it is on).

6. **Press the Starter Switch.** (The large push-button near the centre of the instrument panel). If the engine fails to start at the first attempt, wait a few seconds before pressing the switch again.

The engine should fire and run *without* the use of the petrol priming control, the "Ki-gass" pump or the choke when warm.

Running an engine for 30 minutes, on a day that the tank is not going to be used, is not recommended. The engine will only become half-warm and internal corrosion may result.

16

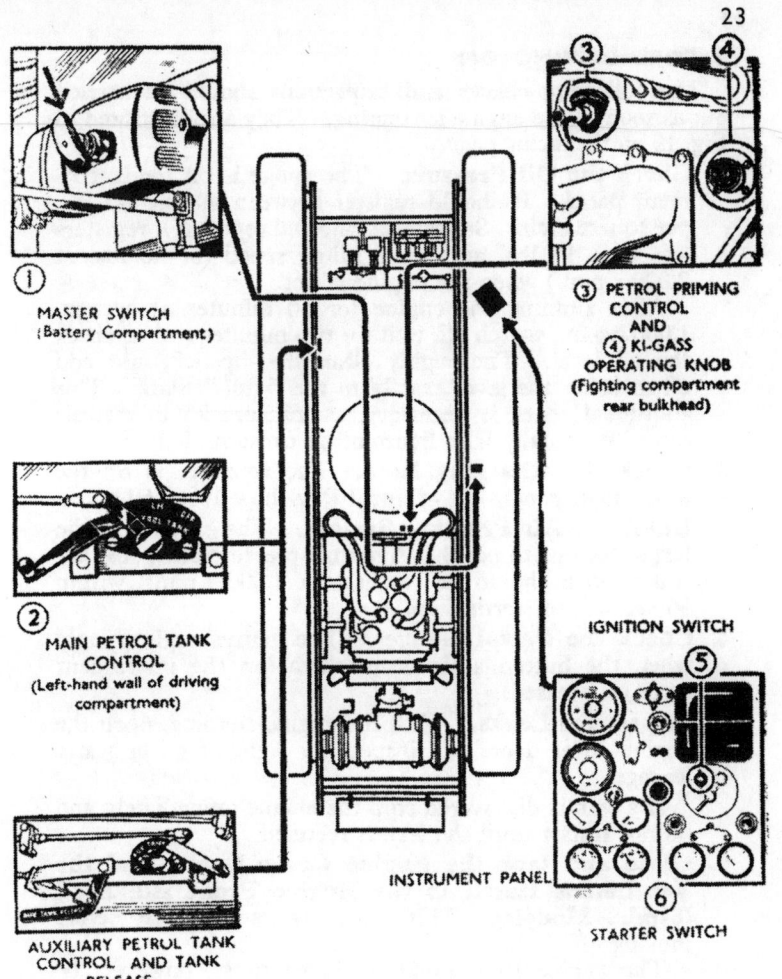

Fig. 12. To start the engine, operate these controls in numerical order. They are explained in the recommended sequence on the facing page.

BEFORE DRIVING OFF

The following checks and inspections should be carried out as soon as the engine is running. They are illustrated in Fig. 13 on the facing page.

1. **Check the Oil Pressure.** The gauge is on the instrument panel. It should register between 40 and 80 lbs. per square inch. Stop the engine and report if it registers less than 15 lbs. at normal idling speed (or 40 lbs. at 2,000 r.p.m.) when the engine is hot.

    After running the engine for 10 minutes at approx. 1,000 r.p.m., switch off, wait for two minutes and re-check the oil level. Thoroughly clean the dipstick, and add oil to bring the level exactly to the " full " mark. This additional check is necessary, as inaccuracies in reading can give a totally false figure of oil consumption.

2. **Check the Charging Rate.** The ammeter is on the instrument panel. It will probably show about 60 amps.

3. **Check the Air Pressure System.** The gauge is to the left of the clutch pedal. A normal pressure of 80 lbs. per square inch should be reached (at 1,000 r.p.m.) within 30 seconds of starting up.

4. **Check the Petrol Gauges.** The gauges register only when the button adjacent to them (on the instrument panel) is pressed.

5. **Inspect for Leaks.** With the engine running, open the engine cover doors and inspect for signs of oil or water leakage.

    If a leak is discovered stop the engine immediately and do not restart until the leak is rectified.

6. **Close and lock the Engine Cover Doors, and the Ventilation Hatch in the Engine Front Bulkhead (6-pdr. Models).** This must always be done before moving off.

    The engine front bulkhead hatch must, however, be open when the gun is being fired.

IMPORTANT.—Do not move the steering handlebar when the vehicle is not being steered. The tank will turn IN NEUTRAL if the bar is operated when the engine is running—see explanation of steering on page 22

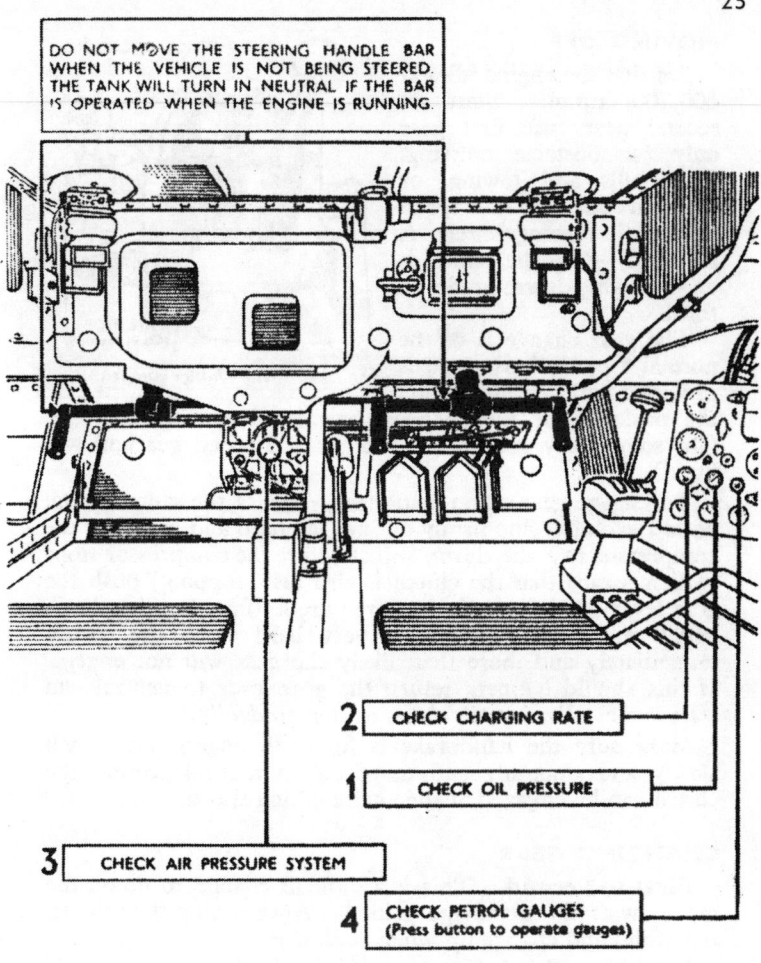

Fig. 13. Check these items when the engine starts, before driving away. Normal readings and other details are explained on the facing page.

MOVING OFF

See that the engine idles at 600/700 r.p.m. Start in second gear (use first gear only for obstacle crossing, freak hills and towing, or when an extremely slow speed or small turning circle is required)—or if the vehicle is on a slight down grade in third.

The gear change is of the normal " crash " type and is provided with a clutch stop;

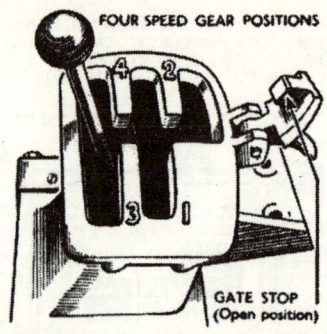

Fig. 14. Gear-change diagrams.

the position for the 4-speed box is shown in Fig. 14. (On some early pre-rework vehicles a 5-speed gearbox was fitted).

To engage gear when stationary, depress the clutch pedal on to the clutch stop firmly but not too hard and listen for the compressor to " die down." Just before the compressor stops (which means that the clutch is also just stopping) push the gear lever firmly into the gear required. Pressing the clutch pedal on to the clutch stop very hard stops the clutch immediately and more than likely the gear will not engage. If this should happen, return the gear lever to neutral and start again, *allowing the clutch to stop gradually.*

Make sure the handbrake is fully off, engage the clutch slowly and smoothly and move off. On level ground the clutch can be engaged at the engine idling speed.

CHANGING GEAR

First to Second. This is a difficult change to do on the move, with no advantage gained. Always stop the vehicle, therefore, and change up when stationary.

Second to Third. Use the double declutch, without throttle or clutch stop, when the vehicle can roll forward at 4 to 5 m.p.h. while the change is being made. Choose a slight, but not steep, down-hill slope if possible. Don't steer while making the change or immediately after it. Rev. up in second gear to 2,200/2,300 r.p.m., take gear into neutral and pull the gear lever gently into third position when the engine revolutions have dropped to 1,500. If the tank stops moving before revs. have dropped sufficiently, start again in second gear. Should the down-hill slope be greater than estimated and the speed increase above 5 m.p.h., engage third gear at 1,700/1,800 r.p.m. for 6 m.p.h., 1,900/2,000 r.p.m. for 7 m.p.h. and so on.

To change to third on the level or uphill, with a rolling speed of less than 5 m.p.h., slow down in second until the tachometer is at 1,500/1,600 r.p.m. Make a fast racing change, using three fingers only, with full use of clutch stop but NO FORCE. Flick the gear lever across from second to third quickly but lightly. Open the throttle immediately third gear is engaged. Never use both hands to make engagement and use the slow double declutch change whenever possible to save gearbox strain. When the engine is temporarily restricted to 2,000 r.p.m., use the slow change only on down-hill slopes and the racing change more often. The engine must not be allowed to rev. up over 2,000 in second gear whilst restricted.

This change is, of course, more difficult when the box is cold and the oil thick, especially in winter. If necessary, run the vehicle for 15 minutes in second gear to warm up.

Third to Fourth. The double declutch method can nearly always be used owing to the higher maximum speed in third gear. Disengage at approx. 2,200 r.p.m. and engage fourth when the revs. have dropped well below 1,500 if the tank is still travelling at 9/10 m.p.h. If the vehicle is accelerating the gear must be engaged at higher revs.

Use the racing change if the ground is slightly uphill. Throttle back in third gear to about 1,600 r.p.m. and push the gear lever straight through into fourth with full depression on the clutch stop.

Fourth to Third. Use the double declutch, with throttle, for all changing down, as on wheeled vehicles. Little throttle is required for this change. Don't stay in top gear if the going is heavy, change to one of the intermediate gears to to ensure ample power for turning.

Third to Second. Change down when the speed is approx. 7 m.p.h. Considerable throttle is required as the engine revs. must be put up to 2,200/2,300 r.p.m. and second gear engaged at just over 5 m.p.h. If the speed is 5 m.p.h. or under, be very quick, with double declutch and less throttle.

Second to First. Normally, stop the tank to engage first gear. If in second gear and pulling up a steep bank, go on until the tank is nearly stationary, dip the clutch quickly, pull straight into first gear and re-engage the clutch quickly. Getting out of second gear after stopping the vehicle on a steep hill is not easy.

Both bottom gear and reverse gear, being very low, tend to become difficult to disengage. Put pressure on the gear lever before dipping the clutch quickly to get into neutral from these low gears, but avoid going on to the clutch stop.

General. Learn the corresponding engine revolutions for each mile per hour in the different gears. Accurate gear changing is quite simple providing the tachometer and speedometer are watched.

Do not change up on slopes steep enough to cause the tank to accelerate and never change down on a down-hill gradient. If a lower gear has not already been engaged for safety reasons on a steep hill and the gradient becomes steeper, turn slowly to the left and stop, get into second gear and re-start.

Owing to the great weight of the vehicle and the engine developing 350 h.p., mistakes in gear changing must lead to serious damage. The gearbox selector forks bend due to the pressure caused by the sliding dogs either grating against the third speed dogs or by the force exerted by the combined leverage and weight of the operating rods when moved by the change speed lever. Jumping out of third gear results from the bending of the third and fourth gear forks. The gearboxes are being progressively improved, but to keep trouble to a minimum, the use of excessive force must be avoided.

STEERING

Move the handlebar firmly and steadily, but without "snatch." Remember that additional effort is needed when operating without compressed air assistance. Always return the bar to the straight ahead position after steering.

Steering in Forward Gears. For forward steering use the steering handlebar in the same way as the handlebar of a bicycle. To turn right pull the right-hand end. To

29

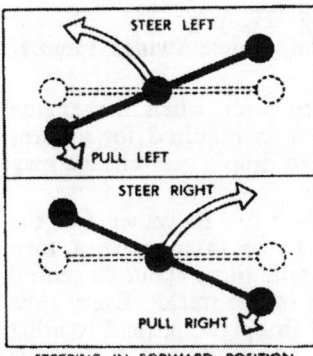

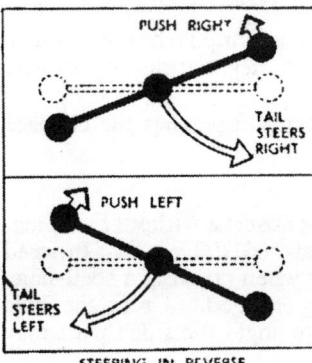

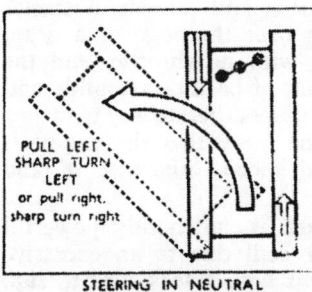

Fig. 15. Pictorial guide to forward, reverse and neutral steering.

turn left, pull the left-hand end.

The turning circle varies with the gear engaged. The smallest turning circle is obtained in first gear and the largest in top gear. (See Turning Radii on page 59.)

Steering in Reverse. To steer in reverse, *reverse the instructions given for forward steering. Push* the right-hand end of the bar to swing the tail to the right. *Push* the left-hand end to swing the tail to the left.

Steering in Neutral. If the steering handlebar is swung while the vehicle is in neutral (engine running) the tracks will rotate in opposite directions and the vehicle will make a sharp pivot turn in little more than its own length.

This is an extremely useful feature but it must not be misused. Never attempt a neutral turn on heavy ground, or where damage to turf is undesirable. And never move the steering handlebar when the engine is running unless a turn is intended. An unintentional pivot turn may have disastrous results.

To carry out a neutral turn, release the handbrake, accelerate *slightly*, and move the steering bar slowly and firmly

23

in the required direction (see Fig. 15, page 23). Do not speed up the engine unduly as the vehicle swings rapidly, especially on a hard surface.

General. Always endeavour to steer when the engine revolutions are high. More power is required for a turn, and if the engine speed is allowed to drop a gear change may be necessary.

Try to complete a turn with one or two movements of the steering bar. Judge which gear to be in to make a turn into any particular opening. The tank turns about its central line and does not follow the front of the track. Every time the bar is operated, power is used in operating (and wearing the linings of) the steering brake. A steady and continuous turn also avoids halting a convoy.

On uneven ground, steer when on a hump rather than when in a hollow. Avoid steering while negotiating an obstacle such as a shell-hole or a river bed.

In general, choose hard rather than soft patches for changes in direction.

OBSTACLE CROSSING

To bring the Churchill over a big obstacle without bumping, make sure that the engine idles at 600/700 r.p.m. Proceed slowly and steadily in bottom gear when crossing a shell-hole. Make sure that the gear is fully engaged. Cross the hole squarely. If it is taken at an acute angle the side thrust may push off one of the tracks.

Approach vertical walls, sleepers and similar obstacles squarely and slowly. In coming over the edge of a drop, allow the tank to creep forward with no throttle, and the vehicle will then ride over the point of balance smoothly and without jar. Do not use the brake unless there is any tendency for the tank to overrun the engine. See that the engine is never driven above the governed speed, otherwise it (and the clutch) will be damaged.

Under all conditions, the engine is sufficiently powerful to slip the tracks and will never stall due to an excessive slope or bank. Continue driving if both tracks start to slip. As the tank digs in, it may get sufficient grip to pull itself out. Failing this, take the machine back and make the run again. If after two or three attempts the tracks continue to slip, make the approach at an angle across the original marks.

BRAKES

Use the brakes intelligently. They are powerful and highly effective, but this fact should not be abused. Don't use them unnecessarily or unnecessarily harshly. Engage a low gear when descending hills to relieve the strain on the brakes. Never allow the vehicle to gain so much momentum on a slope that maximum braking effort is required to control it. Such misuse causes excessive wear, and, in extreme cases, may burn the linings and so render them ineffective.

(*NOTE.*—Take particular care, when using the engine compression as a brake, that the reading on the tachometer does not exceed 2,000 r.p.m.)

DURING HALTS

When the vehicle is halted for short periods, switch off the engine, set the petrol controls to " OFF," turn off the master switch and replace the safety catch on each of the CO_2 fire extinguisher bottles.

HARBOURING AND THE USE OF SIGNALS

If an observer is guiding the tank, which is travelling forward, he raises a closed fist to indicate a turn, using left or right according to the direction required. The driver turns the nose of the vehicle towards the closed hand.

If the tank is in reverse, the observer uses the same signals, but these refer to the tail of the machine and not the front.

For the signal stop, the palm of the hand is held up; for acceleration, the hand is rotated.

When harbouring, remember that the vehicle takes up a lot of space and damage can be done without the driver's knowledge when the tail is swung round. When entering a narrow gateway, always make certain that the whole of the vehicle is through before turning.

AFTER THE RUN

Attend to the following points when the vehicle is parked at the end of the day's run :

Switch off the engine.

Set the petrol controls to " off."

Turn off the master switch.

Refill the petrol tanks.

Check water and oil levels.

Replace safety catch on each of the CO_2 fire extinguisher bottles.

Engage first gear and apply the handbrake.

Check the free travel of the clutch pedal. If it is down to an inch or less the clutch needs adjustment.

Examine the petrol, oil and water connections.

Test by hand the temperature of brake drums, final drive, suspension and bogies.

Inspect tracks and suspension for slackness and damage.

Make sure that final drive spigot bolts and coupling bolts are tight.

STOWAGE—2-PDR.
(Churchill I and II)

Illustrated on Plates A to H—pages 33 to 40.

The following is a complete list of the stowage items carried by the Churchill I and IIR. The numbers in column 1 are the numbers by which the items are identified on the accompanying illustrations. Columns 2 and 3 list, respectively, the articles carried and the number of each supplied. Column 4 contains the reference letters of the illustrations on which the various items can be found. Drawings prepared for an earlier edition have been revised for this issue, necessitating the omission of numbers 47, 48 and 54 in column 1.

VEHICLE EQUIPMENT

Item No.	Description	Number Carried	Plate Ref.
1	Fire extinguisher, CO_2—7 lb. bottle	1	D.
2	Fire extinguisher, Pyrene tetrachloride type—1 quart	2	B. G.
2a	Fire extinguisher—methyl bromide	2	H.
3	Water tanks—drinking	2	B. D.
4	Camouflage net bin :—		H.
	Camouflage net, 35 ft. by 15 ft. ...	1	
	3-piece waterproof cover for vehicle	1	
	Tent poles, 18 ins.	2	
	Tent pegs	4	
	Hemp rope, 20 ft.	1	
	Strips and disc, ground/air communication, per 10 vehicles ...	1 set	
	Matchet (15 in.) in case	1	
5	Flag bag, with 3 signalling flags (1 set)	1	H.
6	Turret headsets (microphone and receiver headgear No. 1, in bag; includes one carried as spare in signal satchel with sling)	4	F.
7	Hull headsets (microphone and receiver headgear No. 2, in bag; includes one carried as spare in signal satchel with sling)	3	A. B.
8	Inspection lamp	1	A.

STOWAGE—2-PDR.—continued

Item No.	Description	Number Carried	Plate Ref.
9	First aid outfit for tanks and armoured cars (packed)	1 box	B.
10	Portable cooker, No. 2 complete	1	C.
11	Miscellaneous bin:—		B.
	Spares and tools—Besa	1 pack	
	Suit—anti-gas (hood, trousers, coat, pair boots and 3 pairs gloves) in bag	1	
	Muzzle cover—Besa	1	
	Muzzle cover—2-pdr.	1	
	*Muzzle cover—3 in. Howitzer	1	
	Breech cover—2-pdr.	1	
	*Breech cover—3 in. Howitzer	1	
	Water bottles	5	
	Tins—biscuit, 10 oz. ration	15	
	Rations, No. 3. 5 men—1 day (packed)	3 boxes	
	Anti-gas ointment, tins of 8 tubes (see also No. 14, Fig. F.)	2	
	Haversacks, W.D. 37 Patt.	2	
	Gloves—wiring	2 pairs	
	Wallet for Bren gun spares	1	
	Pouch—Thompson sub M.G. accessories	1	
	Cotton waste	1 bdle.	
	Cleaning brush, 11 in.	1	
	Brush, bass hand, Mk. I	1	
	* Not required with front Besa gun.		
12	Haversacks, W.D. 37 Patt.	3	A. C.
13	Bleach powder, 2 lbs.	1 tin	G.
14	Anti-gas ointment, tin of 8 tubes (see also No. 11, Fig. B.)	1	F.
15	Anti-dim compound and flannelette	1 jar	F.
16	Shovel, G.S.	2	H.
17	Pickaxe with helve	1	H.
18	Crowbar, 5 ft.	1	H.
19	Sledge hammer (10 lbs.) with handle	1	H.
20	Water cans	4	H.

STOWAGE—2-PDR.—continued

Item No.	Description	Number Carried	Plate Ref.
21	Matchet (15 in.) in case	1	B.
22	Canvas bucket, Mk. V	1	A.
23	Map case, G.S. No. 1, Mk. I	1	E.
24	Wire cutters, folding, Mk. I (in frog)	1	A.
25	Hellesen hand lamp	1	E.
26	Wallet for commander	1	F.
27	Anti-gas cape and gloves	5	A. C.
28	Greatcoats	5	H.
29	Instruction book and tool lists	1	A.
30	Log book (Army Book AB.413)	1	A.
31	Hawser for towing—15 ft.	2	H.
32	Packing blocks for jack (wood)	2	H.
33	Auxiliary battery charging set	1	B.

EQUIPMENT SPARES

34	Prisms, object, Vickers' tank periscope	6	B. E. F.
35	Cleaning brush, prism	3	B. E. F.
36	Vision door triplex unit	2	A.
37	Spare valve bin :—		F.
	Wireless valves, 9—see Parts Tables for details	15	
38	W.T. spare parts case, No. 5 c.	1	F.
	Bulbs, F., 12 v., 2.4 w.	4	
	W.T. key and plug, No. 9	1	
	Rotary transformer brushes (H.T.) No. 4	4	
	Rotary transformer brushes (L.T.) No. 18	2	
	Fuses (radio) 250 m/a	12	
	Holders, No. 1 caps	2	
	Adjuster screws	6	
39	Main wireless aerial	1	B.
40	Troop set wireless aerial	2	F.

VEHICLE TOOLS

41	Vehicle tools (see Schedule of Tools for details)	1 set	A.

STOWAGE—2-PDR.—continued

Item No.	Description	Number Carried	Plate Ref.
42	Hose—bleeding Lockheed system	1	A.
43	Container—bleeding Lockheed fluid	1	A.
44	Lifting jack—hydraulic—10 ton	1	D.
45	Jack handle	1	H.
46	Tecalemit compressor, Junior No. 2, with ball swivel nozzle, type B.S.42	1	B.
49	Spare bulb container	1	A.
	Spare side lamp bulb	4	
	Spare head lamp bulb	1	
	Spare U.V. lamp bulb	1	
50	Idler adjusting packing	70	B.
	(Less those in use)		
51	Bag, spare track pins, No. 1, containing 4 spare pins	1	B.
52	Track link (with 1 in. dia. pin)	2	H.
	Track link pin (includes 4 in bag, see above)	6	
	Track link pin retainer. With vehicle tools (see No. 41, Fig. A.)	6	
	Track link pin retainer lock. With vehicle tools (see No. 41, Fig. A.)	12	
	or		
	Track link (with 1¼ in. dia. pin)	2	H.
	Track link pin (Service) (includes 4 in bag, see above)	6	

ARMAMENT

53	M.G. Bren—.303 Mk. I	1	G.
55	Thompson sub-machine gun	2	B. G.
56	Signal pistol No. 1 Mk. III or Mk. IV	1	E.

ARMAMENT TOOLS

57	Piasaba bin :—		G.
	*Piasaba cleaner No. 20 with rope lanyard and lead ball (3 in. Howitzer)	1	

STOWAGE—2-PDR.—continued

Item No.	Description	Number Carried	Plate Ref.
	Bristle cleaner No. 1, Mk. I, with rope lanyard and lead ball (2-pdr.)	1	
	Cleaning rod—Bren cylinder ...	2	
	Brush, cylinder rod—Besa ...	2	
	*Sponge cap No. 6—3 in. Howitzer (container for Piasaba cleaner)	1	
	Sponge cap No. 4—2-pdr. (container for bristle cleaner) ...	1	
	Protector, object glass, in case. Spares for telescope	6	
	Grease, M.G. Mk. I	2 cans	
	Wire cutters—in frog	1	
	*Not required with front Besa gun.		
58	2-pdr. and *3 in. Howitzer tools and spares (see Parts Tables for details of Tools)	1 box	G.
	*Not required with front Besa gun.		
59	Oil can—recoil replenishment, 2-pdr.	1	E.
60	Besa spares bin :—		B. F.
	Clearing plugs—Besa med. M.G., Mk. I	4	
	Oil can—M.G., Mk. II	1	
61	Cleaning rod, .303 M.G., Mk. IIb	1	G.
62	Cleaning brush, 2 in. bomb thrower	1	E.

ARMAMENT SPARES

63	Striker case complete—spare ...	2	A. E.
64	Sighting telescope No. 30, in case No. 6 (optional)	1	G.
	Sighting telescope No. 33, in case No. 10 (optional)	1	G.

AMMUNITION

65	Cartridge, Q.F., 3 in. Howitzer wh/shell H.E. Mk. II, fuse 119, clip 9 (*Not required with front Besa gun*)	58	A.

STOWAGE—2-PDR.—continued

Item No.	Description	Number Carried	Plate Ref.
66	Q.F., 2-pdr.	150	C. D. G.
67	S.A.H. 29, Mk. I/L expendable (optional)	21 boxes	C. D.
	Belt, Besa med. M.G., Mk. I or II	21 boxes	E. G.
68	Bren .303 M.G. No. 2, Mk. I	6 drums	D. F.
69	Thompson sub-machine gun (one box in position on forward gun). 20 rounds per box	42 boxes	B. F.
70	Bomb, M.L. smoke, 2 in. mortar Mk. I	25	E.
71	Cartridge for signal pistol (8 also carried in transport)	12	F.
72	Hand grenade—Mills	6	F.

AMMUNITION REQUIRED WITH FRONT BESA GUN

73	S.A.H. 29, Mk. I/L expendable (optional)	22 boxes	B.
	Belt, Besa med. M.G., Mk. I or II (optional)	22 boxes	B.

The following items are stowed loose at crew's discretion:—

Cover—bomb thrower	1
Cover—Thompson sub M.G.	2
Cover—Bren muzzle	1
Funnel, collapsible—fuel	1
Filter, collapsible fuel funnel	1
Box—spare maps	1
Ground sheets, Mk. VIII	5
Blankets	5
Cushion seat—spare	1
Case—binoculars	1
Stretcher, ambulance, Mk. II; and slings	2

Fig. A. Churchill I and II. Driver's compartment—right-hand side. Churchill 2-pdr.

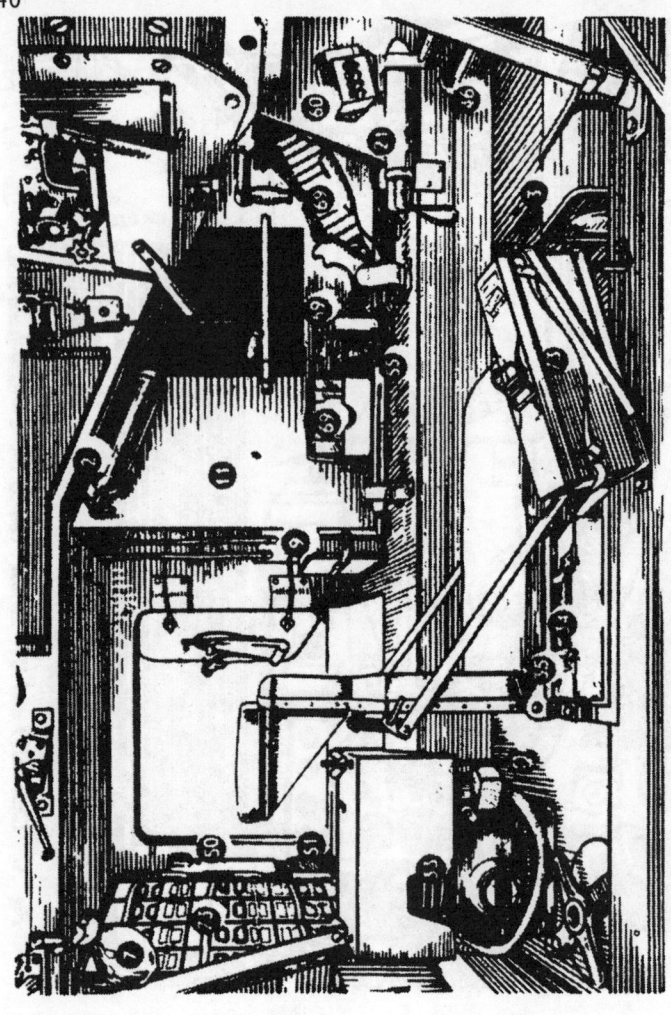

Fig. B. *Churchill I and II. Driver's compartment—left-hand side. Churchill 2-pdr.*

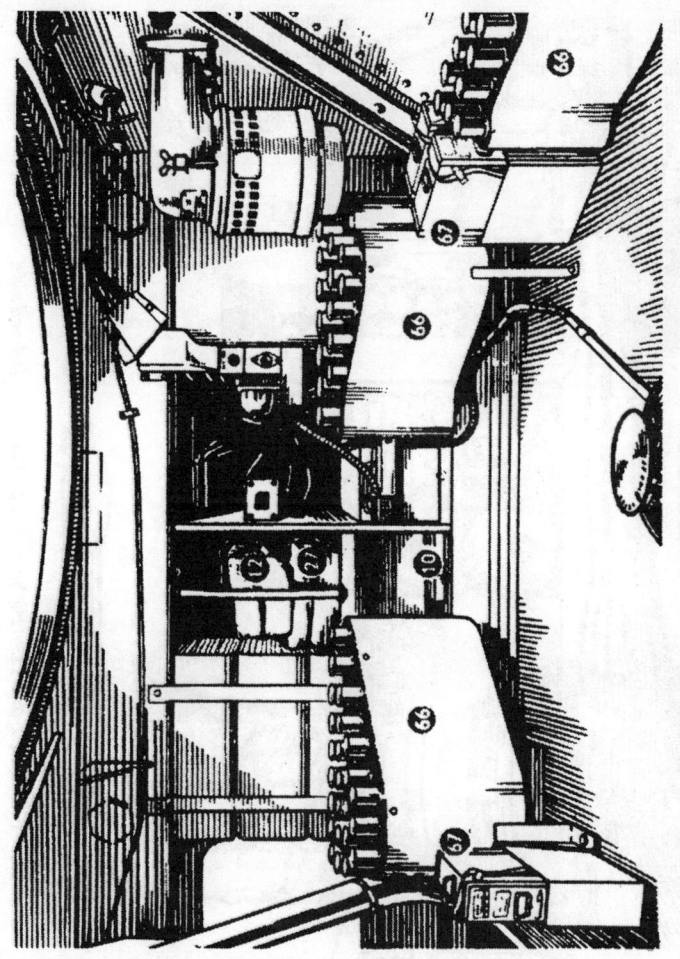

Fig. C. *Churchill I and II. Fighting compartment—right-hand side. Churchill 2-pdr.*

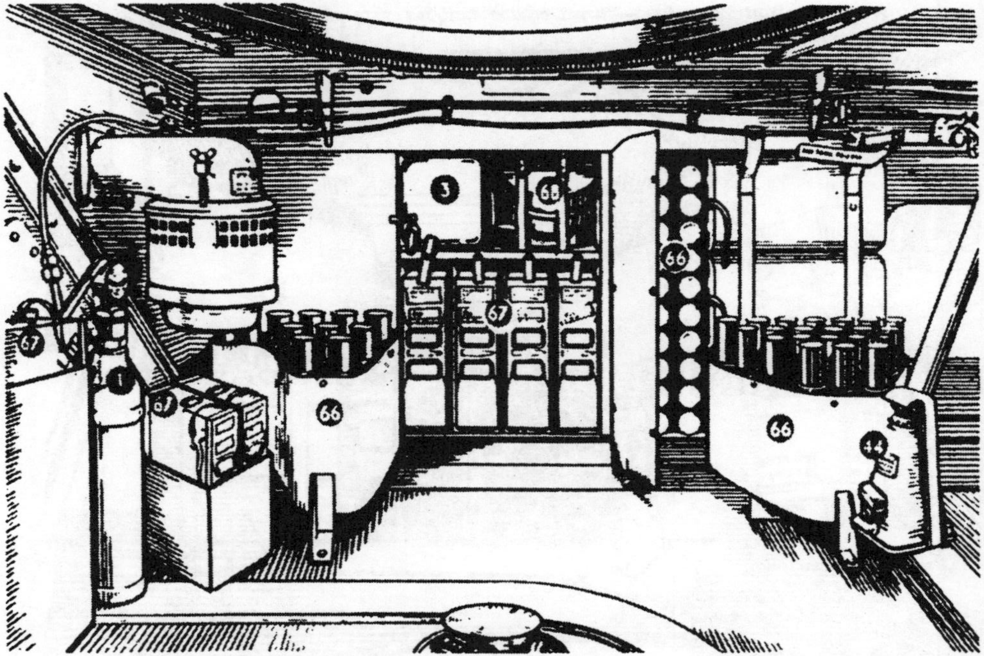

Fig. D. Churchill I and II. Fighting compartment—left-hand side. Churchill 2-pdr.

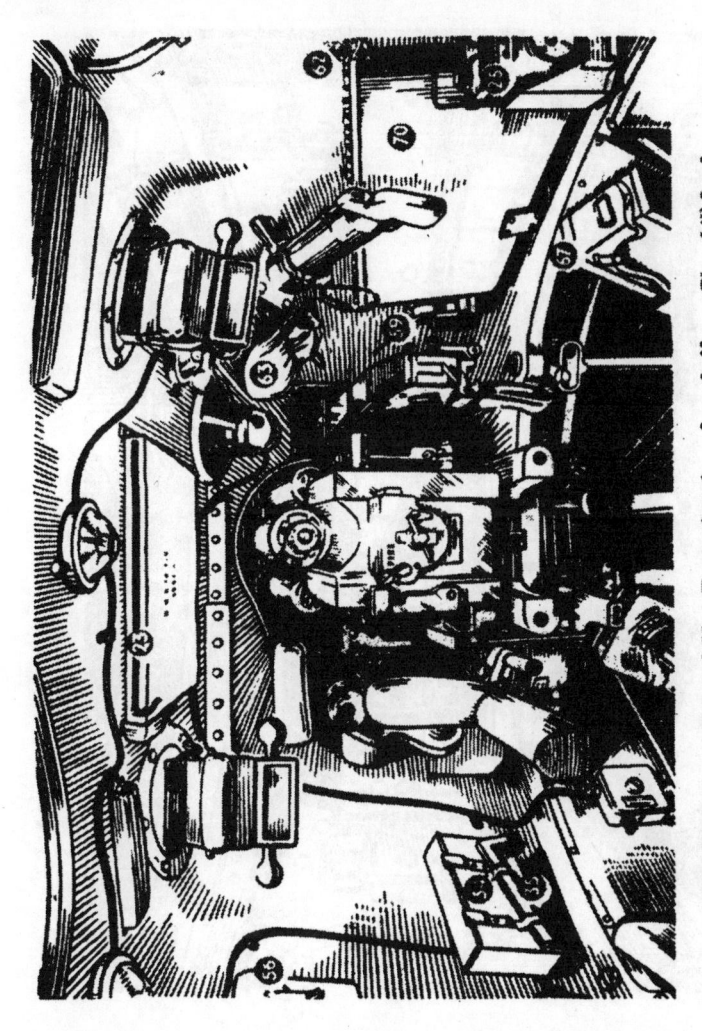

Fig. E. *Churchill I and II. Turret interior—front half. Churchill 2-pdr.*

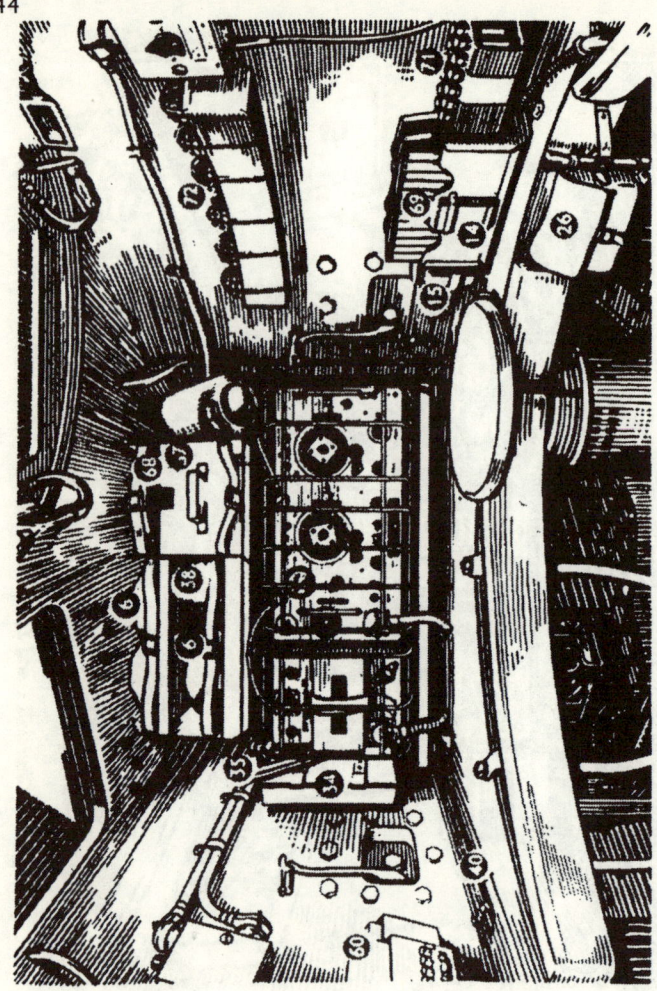

Fig. F. Churchill I and II. Turret interior—rear half. Churchill 2-pdr.

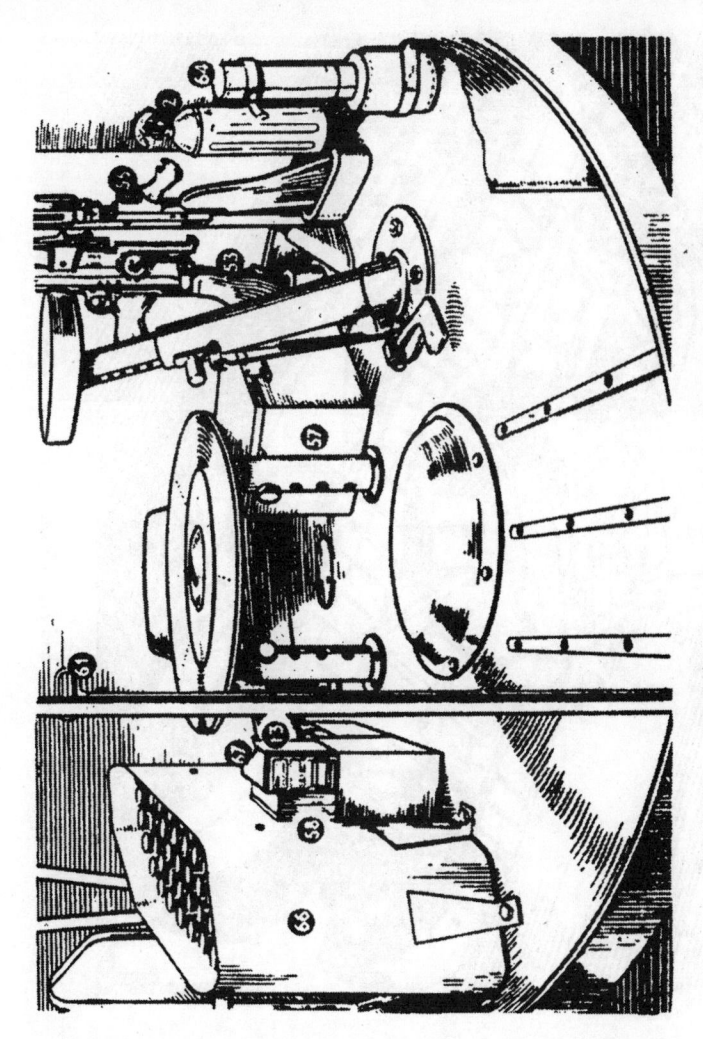

Fig. G. *Churchill I and II. Turret platform. Churchill 2-pdr.*

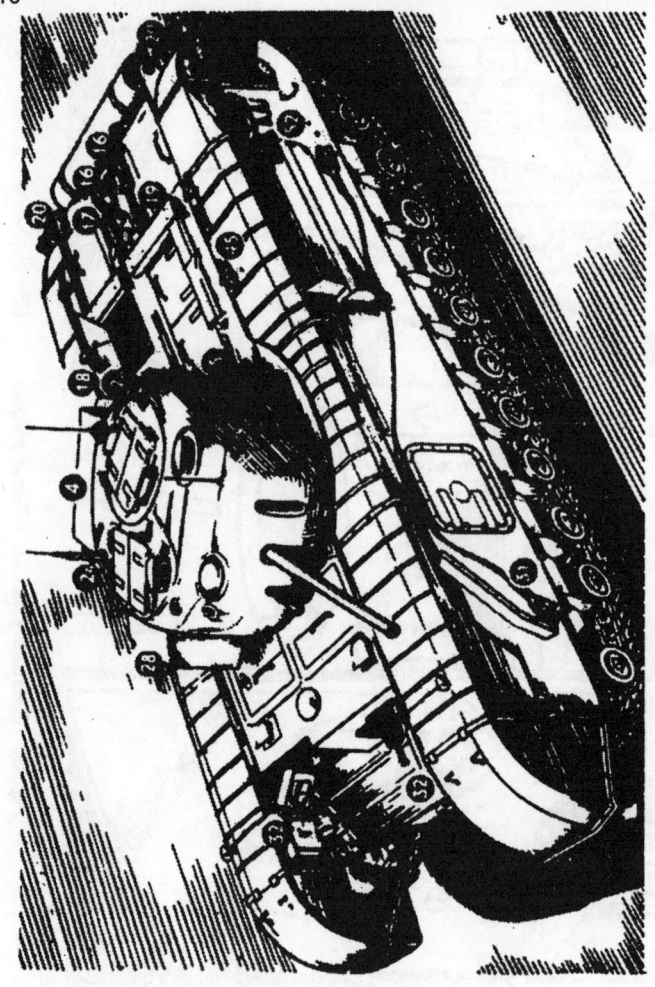

Fig. H. *Churchill I and II. Outside of vehicle. Churchill 2-pdr.*

STOWAGE—6-PDR.
(Churchill III and IV)
Illustrated on Plates I to S—pages 48 to 58.

The following is a complete list of the stowage items carried by the Churchill III and Churchill IV. The numbers in column 1 are the numbers by which the items are identified on the accompanying illustrations. Columns 2 and 3 list, respectively, the articles carried and the number of each supplied. Column 4 contains the reference letters of the illustrations on which the various items can be found. Drawings prepared for an earlier edition of the handbook have been utilised for this issue, necessitating omissions to the numbers in column 1.

VEHICLE EQUIPMENT

Item No.	Description	Number Carried	Plate Ref.
1	Fire extinguisher, CO_2—7 lb. bottle	2	K. L.
2	Fire extinguisher, Pyrene tetrachloride type—1 quart	2	J. O.
2a	Fire extinguisher—methyl bromide	2	P.III. S.IV
3	Water tanks—drinking	2	J. L.
4	Greatcoats and camouflage net bin :—		P.III. S.IV
	Camouflage net for vehicle, 35 ft. by 15 ft.	1	
	3-piece waterproof cover for vehicle	1	
	Tent poles, 18 ins.	2	
	Tent pegs	4	
	Hemp rope, 20 ft.	1	
	Matchet (15 ins.) in case (see also No. 24, Fig. J.)	1	
	Greatcoats	5	
	Strips and disc, ground/air communication, per 10 vehicles	1 set	
5	Flag bag, with 3 signalling flags (1 set)	1	P.III. S.IV

41

STOWAGE—6-PDR.—continued

Item No.	Description	Number Carried	Plate Ref.
6	Turret headsets (microphone and receiver headgear No. 1; includes one carried as spare in signal satchel with sling)	4	N.III. R.IV
7	Hull headsets (microphone and receiver headgear No. 2, in bag; includes one carried as spare in signal satchel with sling)	3	I. J.
8	Inspection lamp	1	I.
9	First aid outfit for tanks and armoured cars (packed)	1 box	J.
10	Portable cooker, No. 2 complete	1	L.
11	Miscellaneous bin :—		J.
	Muzzle cover—Bren M.G.	1	
	Cover—Thompson sub. M.G.	2	
	Pack—spares and tools—Besa	1	
	Gloves—wiring	2 prs.	
	Biscuit, 10 oz. ration	15 tins	
	Rations, No. 3. 5 men—1 day (packed)	3 boxes	
	Water bottles	5	
	Muzzle cover—Besa med. M.G.	2	
	Muzzle cover—6-pdr.	1	
	Cleaning brush, 11 in.	1	
	Anti-gas ointment, tins of 8 tubes (see also No. 15, Fig. M.III or Q.IV.)	2	
	Haversacks—W.D. 37 Patt.	4	
12	Haversack—W.D. 37 Patt.	1	I.
	First aid outfit, small	1 box	S.
14	Bleach powder, 2 lb.	1 tin	N.III. R.IV
15	Anti-gas ointment, tin of 8 tubes (see also No. 11, Fig. J)	1	M.III. Q.IV

42

STOWAGE—6-PDR.—continued

Item No.	Description	Number Carried	Plate Ref.
16	Anti-dim compound, No. 2 (with flannelette)	1 jar	N.III. R.IV
17	Shovel, G.S.	2	P.III. S.IV
18	Pickaxe with helve	1	P.III. S.IV
19	Crowbar, 5 ft.	1	P.III. S.IV
20	Sledge hammer (10 lbs.) with handle	1	P.III. S.IV
22	Recognition signals—tank/air	1	N.III. R.IV
23	Ground sheets, Mk. VIII	5	I.
24	Matchet (15 ins.) in case (see also No. 4, Fig. P.III or S.IV.)	1	J.
25	Canvas bucket, Mk. V	1	I.
	Map board, G.S. No. 2, Mk. 1	1	R.
27	Wire cutters, folding, Mk. I (in frog)	1	I.
28	Hellesen hand lamp	2	L. N.III. R.IV
29	Wallet for commander	1	M.III. R.IV
30	Binoculars, prismatic	1	N.III. R.IV
31	Waterproof suits	2	L.
32	Suit—anti-gas	1	L.
33	Breech cover—6-pdr.	*1	L.
34	Brush, bass hand, Mk. I	1	L.
35	Anti-gas cape and gloves	5	I. L.
36	Instruction book and tool lists	1	I.
37	Log book (Army Book AB.413)	1	I.
38	Hawser for towing, 15 ft.	2	P.III. S.IV
39	Packing blocks for jack (wood)	2	P.III. S.IV
40	Auxiliary battery charging set	1	J.
41	Oil can, ½-pt., with angle spout	1	I.
42	Oil can, 1 gallon, Mk. II/L	1	K.
43	Water cans, 2-gallon	4	P.III. S.IV
45	Cocking lanyard, No. 4, Mk. I	1	M.III. Q.IV

EQUIPMENT SPARES

46	Prisms, object, Vickers' tank periscope.	12	J. M.III. N.III. Q.IV. R.IV

43

STOWAGE—6-PDR.—continued

Item No.	Description	Number Carried	Plate Ref.
47	Cleaning brush—prism	3	J. M.III. N.III. Q.IV. R.IV
48	Vision door triplex unit	2	I.
49	Spare valves case	1	N.III. R.IV
	Wireless valves. (See Parts Tables for details)	15	
50	W.T. spare parts case No. 5c	1	N.III. R.IV
	Bulbs, F., 12 v. 2.4 w.	4	
	W.T. key and plug, No. 9	1	
	Rotary transformer brushes (H.T.), No. 4	4	
	Rotary transformer brushes (L.T.), No. 18	2	
	Fuses (radio), 250 m/a	12	
	Holders, No. 1 caps	2	
	Adjuster screws	6	
51	Main wireless aerial	1	J.
52	Troop set wireless aerial	2	N.III. R.IV

VEHICLE TOOLS

Item	Description	No.	Plate
53	Vehicle tools, set (see Schedule of Tools for details)	1	I.
54	Hammer, engineer's, 2 lb. (Part of No. 53)	1	I.
55	Hose—bleeding Lockheed system	1	I.
56	Container—bleeding Lockheed fluid	1	I.
59	Tecalemit compressor, Junior No. 2, with ball swivel nozzle, type B.S.42	1	J.

VEHICLE SPARES

62	Spare bulb container	1	I.
	Spare side lamp bulb		

44

STOWAGE—6-PDR.—continued

Item No.	Description	Number Carried	Plate Ref.
	Spare head lamp bulb ...	1	
	Spare U.V. lamp bulb ...	1	
63	Idler adjusting packing ...	70 (Less those in use)	J.
64	Track link (with 1 in. dia. pin)	2	P.III. S.IV
	Track link pin, includes 4 in bag, (see No. 65, Fig. L.)	6	
	Track link pin retainer. With vehicle tools (see No. 53, Fig. I.)	6	
	Track link pin retainer lock. With vehicle tools (see No. 53, Fig. I.)	12	
	or		
	Track link (with 1¼ in. dia. pin)	2	
	Track link pin (Service) (includes 4 in bag, see No. 65, Fig. L.)	6	
65	Bag, spare track pins, No. 1, containing 4 spare pins ...	*1	L.

ARMAMENT

66	M.G. Bren .303 Mk. I or Mk. II	1	O.
68	Thompson sub-machine gun	2	J. O.
69	Signal pistol No. 1, Mk. III or Mk. IV	1	N.III. R.IV

ARMAMENT TOOLS

70	Piasaba bin :—		O.
	Bristle cleaner, 6-pdr. ...	1	
	Sponge cap, 6-pdr.	1	
	Brush, cylinder rod ...	2	
	Cleaning rod—Bren cylinder	1	
	Protector, object glass ; in case. Spares for telescope	6	

STOWAGE—6-PDR.—continued

Item No.	Description	Number Carried	Plate Ref.
	Wire cutters, folding, Mk. I (in frog)	1	
72	Wallet—Bren .303 M.G. Mk. I spares	1	L.
73	Gun tool bin :—		K.
	Tools and spares, etc., 6-pdr. (see Parts Tables for details)	1	
	Cleaning rod—Bren cylinder	1	
74	Cotton waste	1 bdle.	N.III. R.IV
75	Oil can—recoil replenishment ..	1	M.III. Q.IV
76	Besa spares bin :—		J. M.III. R.IV
	Clearing plugs—Besa med. M.G., Mk. I	4	
	Oil can, M.G., Mk. II ...	1	
77	Cleaning rod—.303 M.G., Mk. IIb or Mk. IV	1	O.
77a	Cleaning rod—.303 M.G., Mk. IIb or Mk. I	1	M.III. Q.IV. R.IV
78	Cleaning brush—2 in. bomb thrower	1	M.III. Q.IV
79	Grease, M.G. Mk. I	1 can	M.III. Q.IV
80	Case and striker—6-pdr. ...	1	N.III. R.IV
81	Sighting telescope No. 33 in case No. 10 (optional) ...	1	O.
	Sighting telescope No. 30 in case No. 6 (optional) ...	1	O.
	Sighting telescope No. 39 in case (optional)	1	O.

AMMUNITION

Item No.	Description	Number Carried	Plate Ref.
82	Cartridge, 6-pdr., Q.F., anti-tank	85	K. L. O.
83	S.A.H.29, Mk. I/L, expendable (optional)	31 boxes	J. K. L. M.
	Belt, Besa med. M.G., Mk. I or II (optional)	31 boxes	N.III. R.IV

STOWAGE—6-PDR.—continued

Item No.	Description	Number Carried	Plate Ref.
84	Thompson sub-machine gun. (One box in position on forward gun) 20 rounds per box	42 boxes	J. M.III. Q.IV
85	Bren, .303 M.G. No. 2, Mk. I	6 drums	L. N.III. R.IV
86	Bomb, M.L. smoke, 2 in. mortar Mk. I	30	M.III. Q.IV
87	Hand grenade (Mills)	12	N.III. R.IV
	Grenade, W.P. Smoke (optional)	12	
88	Cartridge for signal pistol	12	N.III. R.IV

The following items are stowed loose at crew's discretion:—

Cover, bomb thrower	...	1
Blankets	...	5
Funnel—collapsible fuel	...	1
Filter, collapsible fuel funnel		1
Cushion seat—spare	...	1
Stretcher, ambulance, Mk. II; and slings	...	2

Fig. I. Churchill III and IV. Driver's compartment—right-hand side. Churchill 6-pdr.

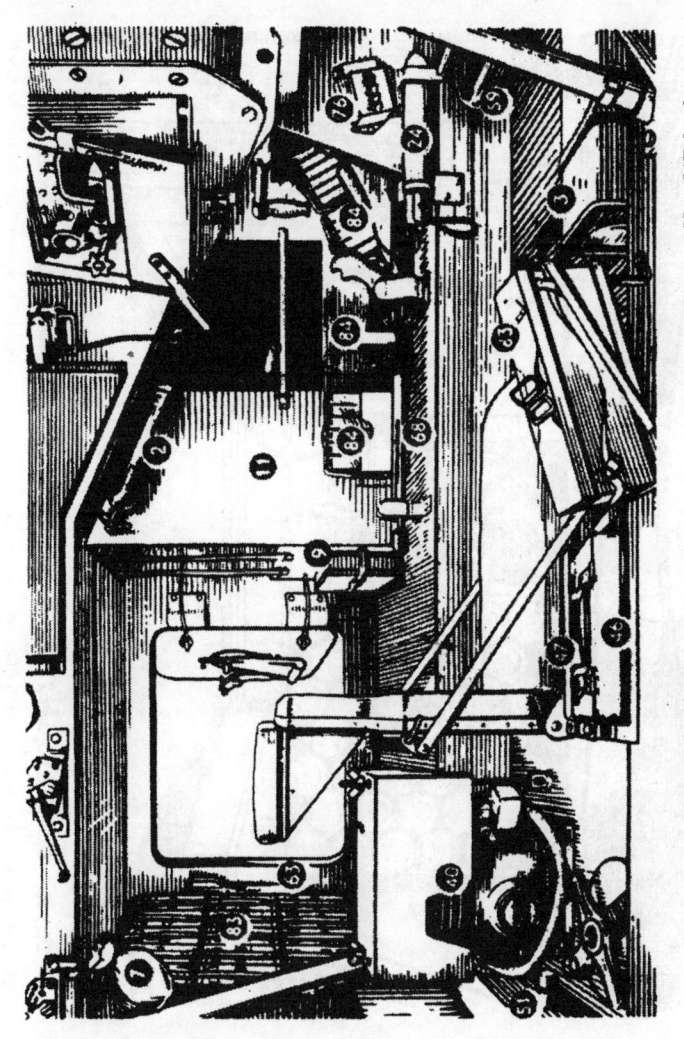

Fig. J. *Churchill III and IV. Driver's compartment—left-hand side. Churchill 6-pdr.*

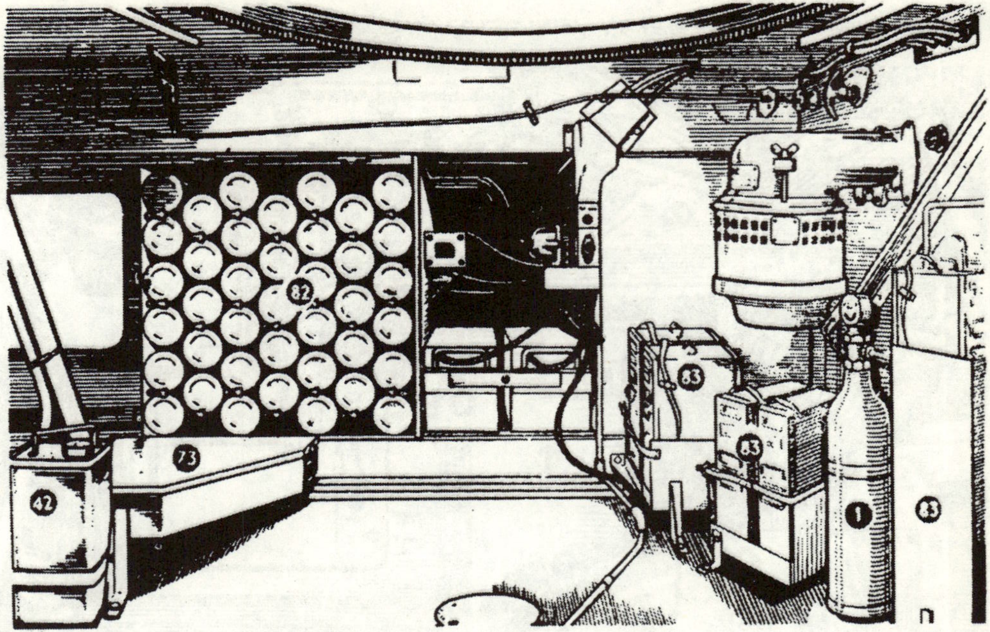

Fig. K. *Churchill III and IV. Fighting compartment—right-hand side. Churchill 6-pdr.*

Slight changes have been made since this illustration was produced. On current vehicles "two-stage" air cleaners are fitted and an accessories bin displaces items 83 in the rear corner, these items now being stowed as noted against Fig.L.

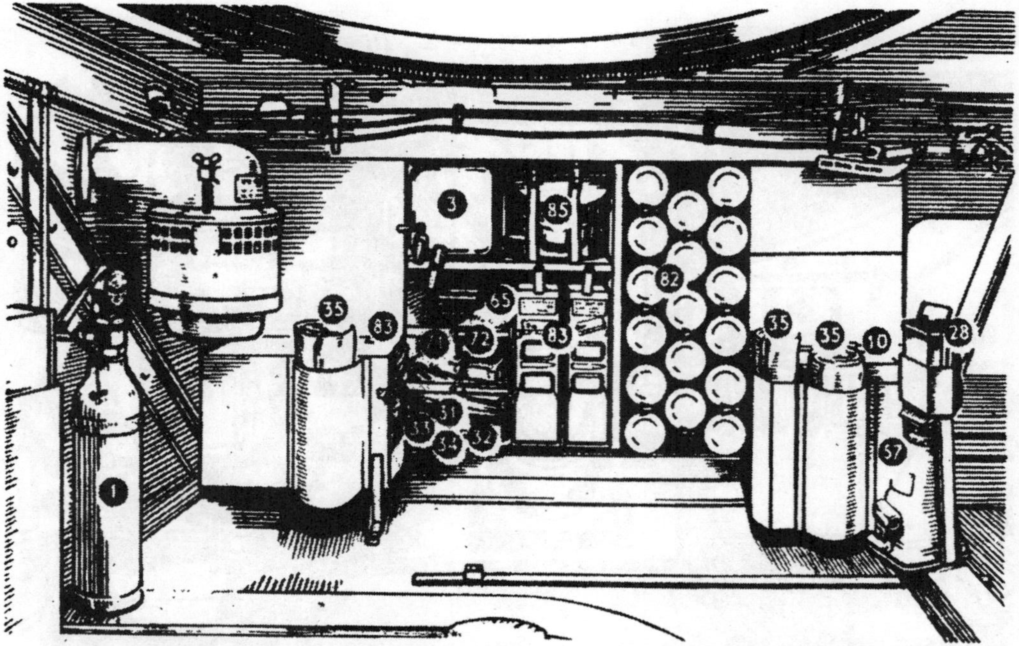

Fig. L. *Churchill III and IV. Fighting compartment—left-hand side. Churchill 6-pdr.*

Sight changes have been made since this illustration was produced. Current vehicles have "two-stage" air cleaners; items 31 to 35, 63 and 72 are stowed at crew's discretion in an accessories bin fitted in the rear corner; items 83 are now together at left-centre. Item 71 is deleted.

51

Fig. M. Churchill III. Turret interior—front half. Churchill 6-pdr.

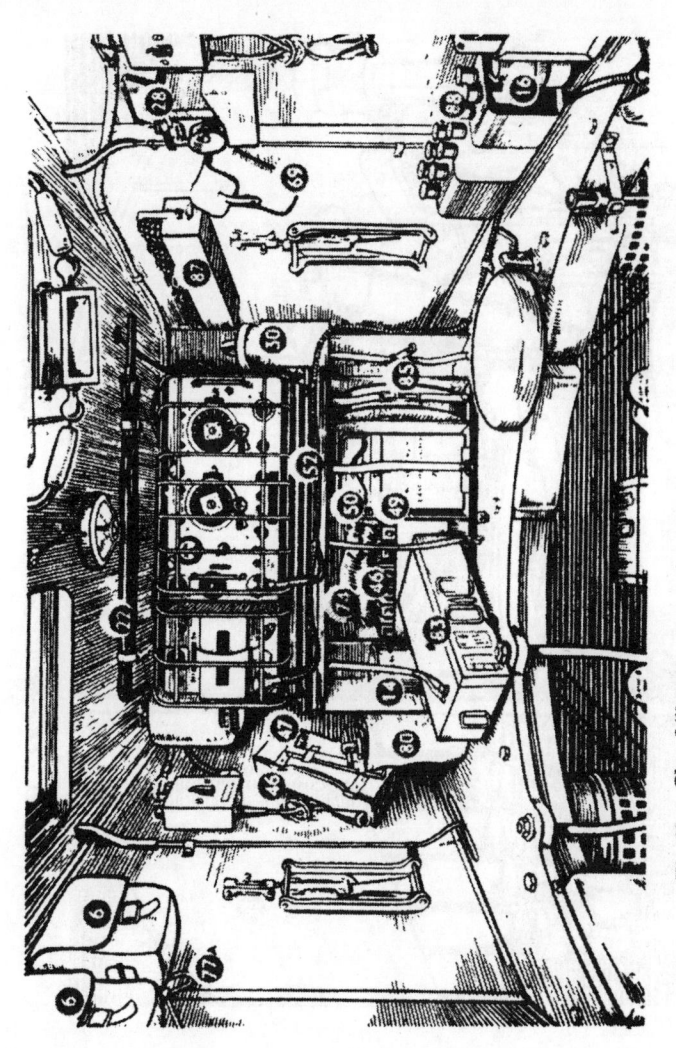

Fig. N. Churchill III. Turret interior—rear half. Churchill 6-pdr.

Fig. O. Churchill III and IV. Turret platform. Churchill 6-pdr.

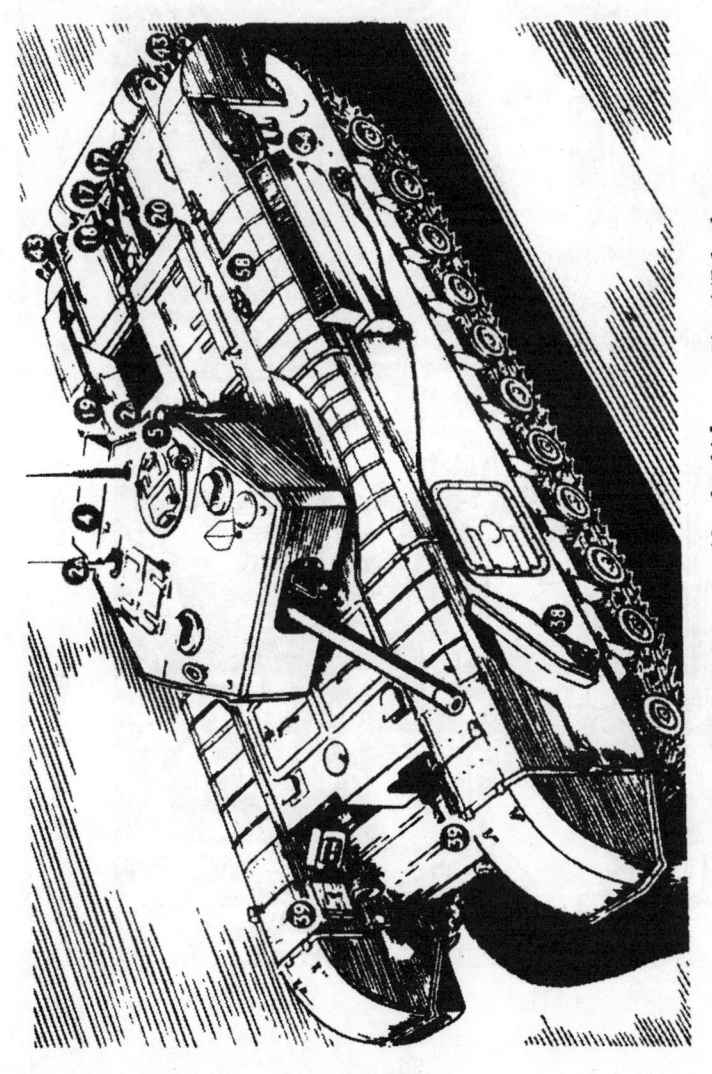

Fig. P. Churchill III. Outside of vehicle. Churchill 6-pdr.

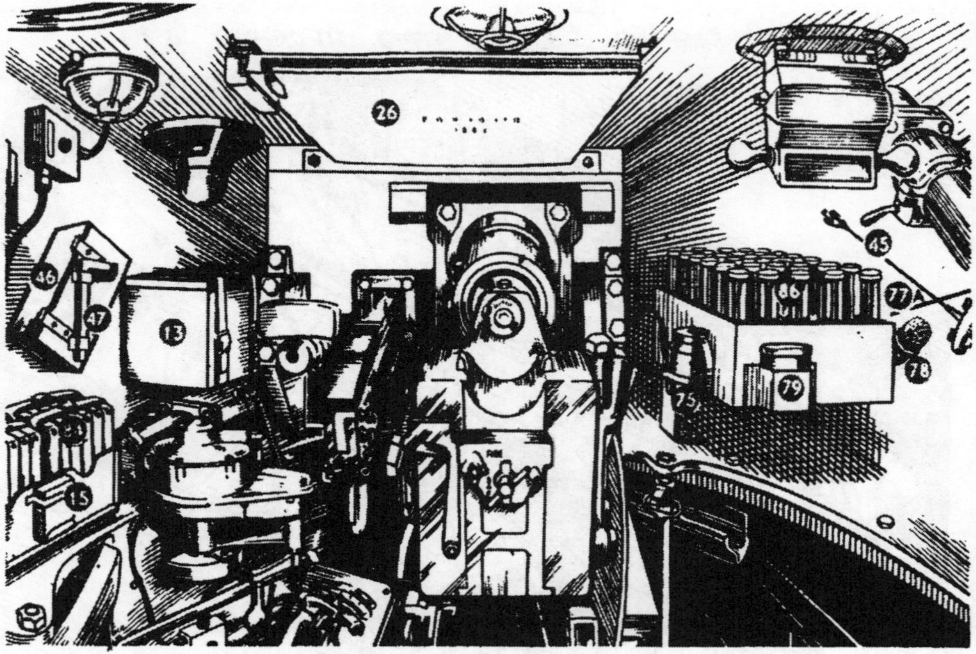

Fig. Q. Churchill IV. Turret interior—front half. Churchill 6-pdr.

Slight changes have been made since this illustration was produced. Items 13 and 26 are now deleted and two hand grenade boxes are added to the right of item 78.

Fig. R. *Churchill IV. Turret interior—rear half. Churchill 6-pdr.*

Slight changes have been made since this illustration was produced. A map board is now placed over the wireless set and the container of item 87 has been altered to make two boxes.

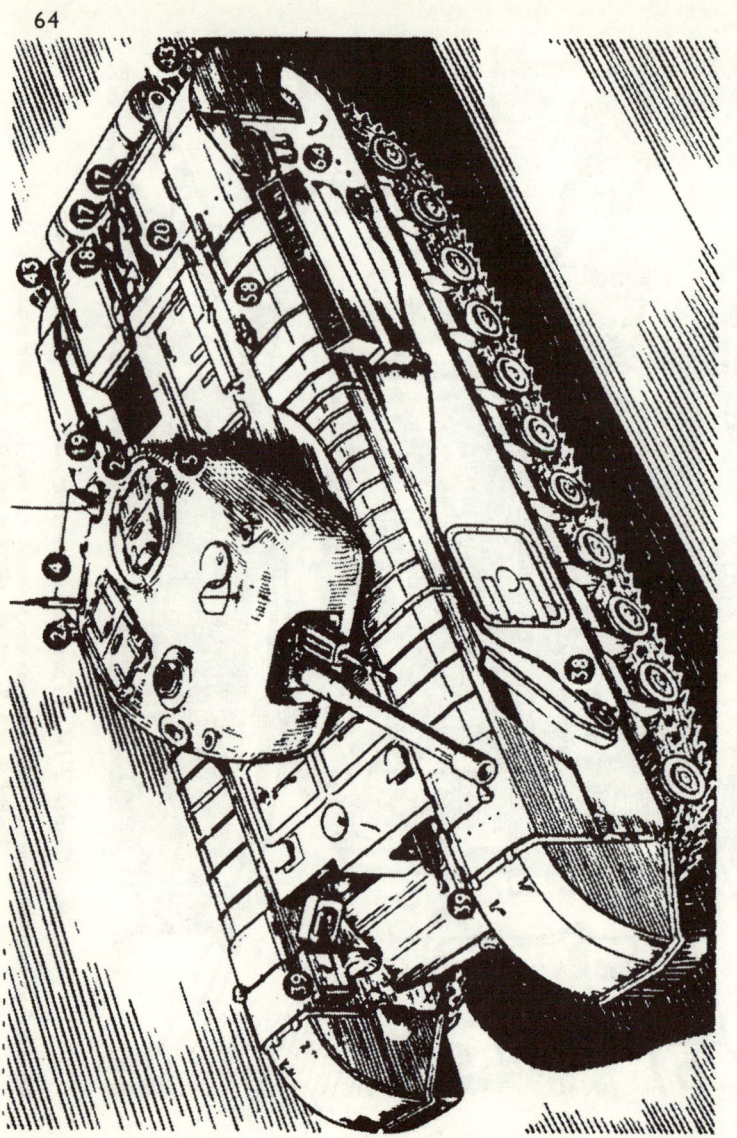

Fig. 5. Churchill IV. Outside of vehicle. Churchill 6-pdr.

Slight changes have been made since this illustration was produced. Items 39 and 58 are deleted and a small First Aid box is added at the rear.

DIMENSIONS AND CAPACITIES

Dimensions

Overall Length of Vehicle	25' 2"
Overall Width of Vehicle	10' 8"
Width without Air Louvres	9' 2"
Overall Height	8' 2"
Ground Clearance	1' 8"
Length of Track on Ground	12' 6"
Width across Tracks	9' 1"

Capacities

Petrol (all Tanks)	182½ gallons approx.
Right-hand Tanks	75 gallons approx.
Left-hand Tanks	75 gallons approx.
Jettison Tank	32½ gallons approx.
Cooling System	26 gallons approx.
	(13 gallons each side)
Oil in Engine	11 gallons approx.
Oil in Gearbox—Dry Box	3 gallons
Refill after draining	2¼ gallons
Oil in each Final Drive Unit	1¾ gallons
Oil in Air Compressor Sump	1½ pints

Turning Radii

First Speed	10.85 ft.
Second Speed	29.8 ft.
Third Speed	57.4 ft.
Fourth Speed	95.9 ft.

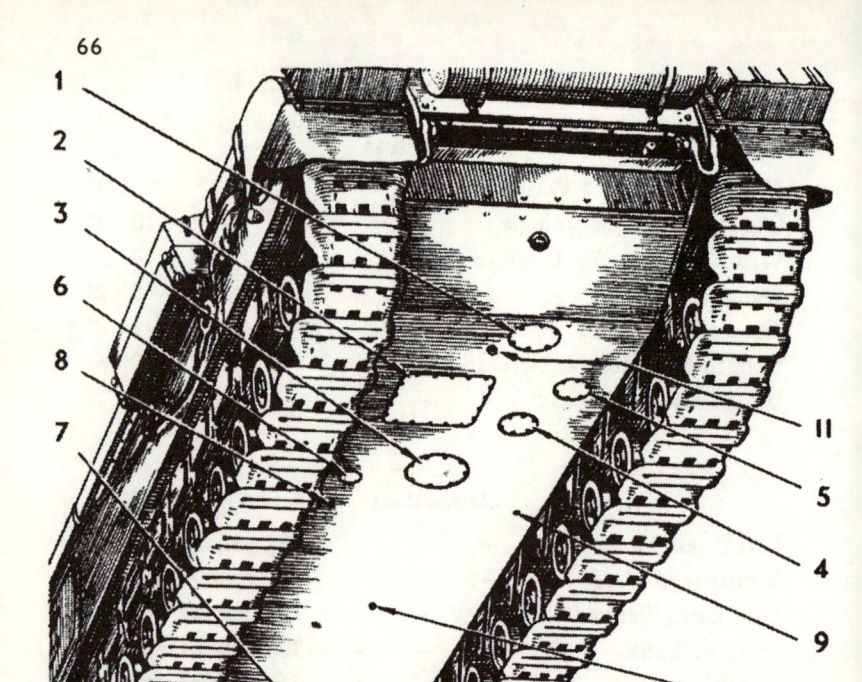

Fig. 16. Inspection plates and traps in hull floor.

Plate or
Trap No.   For Access to

1 Gearbox drain plug. Clutch control pipe unions.
2 Starter motor. Centre sump rear cover plate (early models). Scavenge pump and pipes. Petrol pump and (where fitted) flexible drive. Bottom hose and fittings of L.H. radiator. 2-way petrol tap and unions.
3 Engine front mounting bolt. Centre sump front cover plate (early models). Scavenge pump suction pipe flange. Centre sump drain plug (where fitted). Bottom connections of water pumps. 3-way petrol tap and unions.
4 Main oil sump drain plug. Petrol system drain plug (in distribution box). Bottom hose and fittings of R.H. radiator.
5 Not used.
6 Petrol release valve (see page 11).
7 Emergency disposal hatch (see page 11).
8 & 9 Drain holes for L.H. and R.H. water systems.
10 Drain plug for rotary base junction.
11 Hull floor drain plug, engine and gearbox compartment.

MAINTENANCE

An athlete is kept at the top of his form by correct feeding and regular exercise. A tank is kept in fighting trim by correct lubrication and regular adjustments. Both may fail at the critical moment if these preliminaries are neglected—but there the comparison ends.

The athlete can fall out of a race with nothing more serious than a slight loss of reputation. The fighting vehicle dare not fail in the middle of an action.

Maintenance, therefore, is of vital importance. It is a wearisome job, a repetition time and time again of the same routine. But it is the surest, safest, wisest form of insurance known to mechanical warfare.

In the following pages the maintenance routine for the Churchill Tank is detailed under periods showing when the respective operations should be carried out. " Lubrication " items are printed in ordinary type, and " Inspections and Adjustments " are shown in capital letters. A lubrication chart is folded inside the back cover.

Make up your mind to follow these instructions conscientiously. Remember, maintenance is meaningless without regularity.

OILS AND GREASE

The correct grades of oil and grease are specified in this handbook and in the instruction book. Alternatives are *not* suitable and should not be used.

For the engine, and for all points requiring engine oil, use 10 H.D.

For the gearbox and all points requiring gear oil, use C.600. (This includes all *pressure nipples* lubricated with the oil gun, and is the only lubricant used in the gun).

For all *screw-down greasers*, use Grease No. 3 (high melting point).

For the Lockheed systems use Lockheed Racing Green or Hydraulic Brake Fluid No. 4. Never use oil (which causes rapid deterioration of the rubber pistons) or an incorrect brake fluid (which may vaporise at the high temperatures generated and thus render the brakes inoperative).

DAILY ROUTINE

1. Check engine oil level. Top up as required with engine oil (10HD). The dip-stick is under (or next to) the filler, and the oil should be replenished up to the " full " mark. (The " low " mark indicates the danger level). The vehicle should be on level ground.

Run engine for 10 minutes at 1,000 r.p.m., switch off, wait for two minutes and check oil level again, with dipstick perfectly clean. Add oil if necessary to bring level *exactly* to the " full " mark.
(*Job No. A.1 in Instruction Book*)

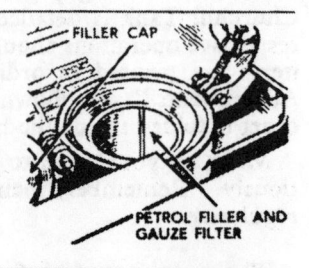

2. Check petrol tank levels. Do not fill main tanks above 4½ in. from top of filler. The auxiliary tank should be filled to the top. Petrol not below 75 Octane must be used. Check also the operation of the petrol controls.
(*Job No. A.1 in Instruction Book*)

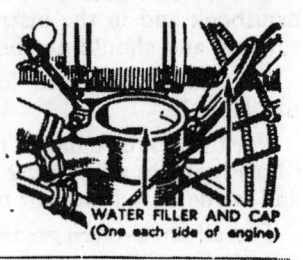

3. Check water levels—one filler cap on each side of engine compartment at the front. Securely fasten both caps after checking as the system operates under pressure.
(*Job No. A.1 in Instruction Book*)

---

4. CHECK CLUTCH OPERATION AT END OF DAY'S RUN FOR CORRECT CLEARANCE AND COMPLETE DISENGAGEMENT, AND ADJUST IF NECESSARY.
(*Job No. A.10 in Instruction Book*)

LUBRICATION OPERATIONS ARE IN ORDINARY TYPE ;

DAILY ROUTINE—continued

5. INSPECT OIL AND FUEL PIPES FOR LEAKAGE OR FRACTURE AT END OF DAY'S RUN.

6. Lubricate bogie axles and fulcrum shafts (22 nipples each) with gear oil (C.600) using pressure gun. Inject oil in bogie axles until surplus flows from vent on rear side of axle shaft bracket, and in fulcrum shafts until oil exudes round the boss of the shaft.
NOTE. Every 50 *miles* is sufficient when *daily* mileage is less than 50.
(*Job No. A.2 in Instruction Book*)

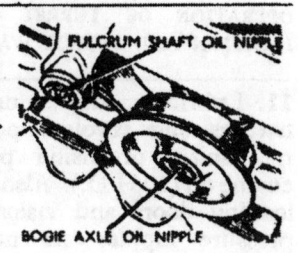

7. INSPECT TRACK SOLE PLATES FOR CRACKS, AND INSPECT FOR LOOSE TRACK PIN RETAINERS. REPORT ANY DEFECTS.

8. INSPECT BOLTS ON BOGIE BRACKET TIE PLATES, AND TIGHTEN IF NECESSARY.
NOTE.—THE REAR BOLT ON EACH PLATE IS ALLOWED A SMALL END CLEARANCE, BUT THE FRONT BOLT SHOULD CLAMP THE TIE PLATE FIRMLY. BOTH NUTS ARE SECURED BY LOCK-NUTS.

9. Check oil level in compressor. Top up as required with engine oil (10 H.D.). Fill to top of filler hole.
(*Job No. A.2 in Instruction Book*)

INSPECTIONS AND ADJUSTMENTS IN CAPITAL LETTERS

DAILY ROUTINE—continued

10. CHECK OPERATION OF ELECTRICAL EQUIPMENT. MAKE SURE THAT THE MAIN DYNAMO IS CHARGING CORRECTLY. CHECK OPERATION OF TURRET TRAVERSE GEAR, MASTER SWITCH, IGNITION AND PETROL WARNING LIGHTS.

---

11. Lubricate hinges on escape hatches and revolver ports, and mechanism of vision port with engine oil (10 H.D.). Also hinges of loading doors and vision port if pressure nipples are not fitted. Lubricate edges of loading doors, vision port, revolver ports and escape hatches with gear oil (C.600).
(*Job No. A.2 in Instruction Book*)

See item 24 on page 71 (250 miles) for hinges fitted with pressure nipples.

---

12. Lubricate gun depression roller shaft (on elevating bracket under gun mounting). One hole each end (C.600).
(*Job No. A.2 in Instruction Book*)

---

13. Check oil level in auxiliary generator engine (whenever generator has been in use). Top up as required with engine oil (10 H.D.). Fill to bottom thread of filler plug.
(*Job No. A.2 in Instruction Book*)

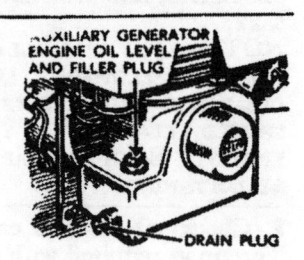

AUXILIARY GENERATOR ENGINE OIL LEVEL AND FILLER PLUG

DRAIN PLUG

LUBRICATION OPERATIONS ARE IN ORDINARY TYPE;

EFFECTS OF OPERATING CONDITIONS

The following extra routine must be carried out at the stated intervals when operating in dusty conditions.

DAILY

On early vehicles wash out air cleaner elements and casings and refill to level mark with engine oil (10 H.D.).

Refer to illustration on page 72 (item 2—upper).

Later vehicles have "two-stage" air cleaners. Remove the bottom dust containers and empty them. Before replacing make sure that the dust ejection slots are free and that no oil drips into the centrifugal conical pans.

Refer to illustration on page 72 (item 2—lower).

(*Job No. A.3 in Instruction Book*)

REMOVE AIR INLET LOUVRES AND CLEAN AIR PASSAGES IN RADIATOR CORES.

IF POSSIBLE, THE AIR PASSAGES SHOULD BE BLOWN OUT WITH COMPRESSED AIR FROM THE CENTRE OF THE VEHICLE OUTWARDS.

THIS WILL NOT BE NECESSARY ON VEHICLES FITTED WITH FULL-LENGTH MUDGUARDS AND UPTURNED LOUVRES.

EVERY 250 MILES

WASH FILTER CLOTH OF AIR COMPRESSOR INLET.

(*Job No. B.21 in Instruction Book*)

Refer to illustration on page 75 (item 16).

INSPECTIONS AND ADJUSTMENTS IN CAPITAL LETTERS

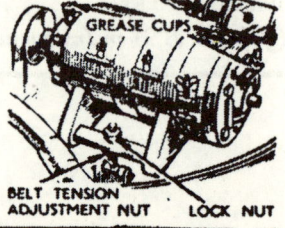

EVERY 250 MILES
(Or WEEKLY when the weekly mileage is less than 250)

1. REMOVE AND EMPTY THE BOTTOM DUST CONTAINERS ON THE "TWO-STAGE" AIR CLEANERS FITTED TO LATER VEHICLES. BEFORE REPLACING THEM, MAKE SURE THAT THE DUST EJECTION SLOTS ARE FREE AND THAT NO OIL DRIPS INTO THE CENTRIFUGAL CONICAL PANS.

Refer to illustration on page 72 (item 2—lower).

2. CHECK TENSION OF WATER PUMP DRIVING BELTS AND ADJUST IF NECESSARY.
(Job No. A.6 in Instruction Book)

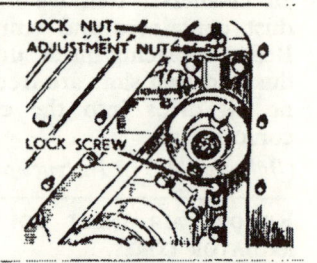

3. Lubricate throttle control rod joints with engine oil (10 H.D.).
(Job No. A.2 in Instruction Book)

4. CHECK THROTTLE CONTROL ROD NUTS FOR TIGHTNESS.

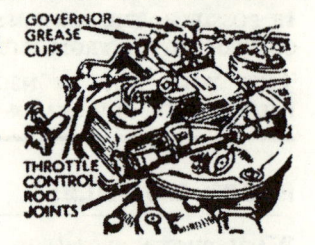

5. CHECK TENSION OF MAIN DYNAMO DRIVING BELT AND ADJUST IF NECESSARY.
(Job No. A.6 in Instruction Book)

LUBRICATION OPERATIONS ARE IN ORDINARY TYPE;

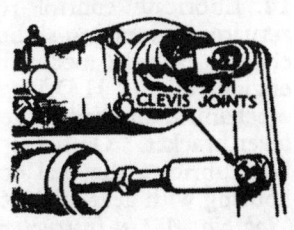

EVERY 250 MILES—continued

6. Examine level of electrolyte in batteries and top up if required. The two 6-volt batteries are fitted on the floor of the battery recess. *Check more frequently in exceptionally hot conditions.*

7. CHECK BATTERY CONNECTIONS FOR TIGHTNESS. CLEAN TERMINALS AND SMEAR WITH VASELINE.

8. Check gearbox oil level of turret traverse gear. Top up if necessary with 10 H.D. to level of filler plug.

(*Job No. A.2 in Instruction Book*)

9. Lubricate three nipples of clutch throw-out mechanism with gear oil (C.600).

(*Job No. A.2 in Instruction Book*)

10. Lubricate clevis joints of clutch servo motor with engine oil (10 H.D.).

(*Job No. A.2 in Instruction Book*)

INSPECTIONS AND ADJUSTMENTS IN CAPITAL LETTERS

EVERY 250 MILES—continued

11. CHECK TENSION OF TURRET GENERATOR DRIVING BELT AND ADJUST IF NECESSARY.
(*Job No. A.6 in Instruction Book*)

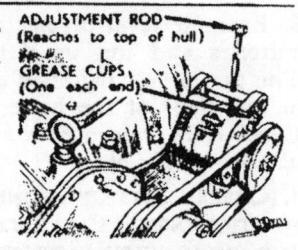

12. Top up fluid level in brake fluid reservoir, if required, *with Lockheed Hydraulic Racing Green or Hydraulic Brake Fluid No.* 4. The filler plug is situated under the driver's escape hatch. The level plug is fitted in the pipe about halfway between the filler plug and the reservoir.
(*Job No. A.2 in Instruction Book*)

13. Lubricate bracket bearings for speedometer drive with engine oil (10 H.D.). Not on vehicles fitted with fixed pulley.
(*Job No. A.2 in Instruction Book*)

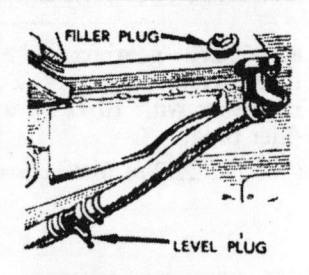

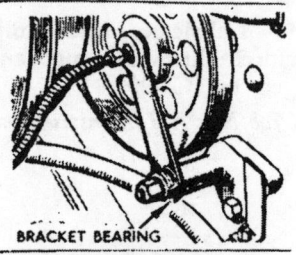

14. Lubricate control rod joints, actuating levers and sliding rods of change speed mechanism with engine oil (10 H.D.). Also gate mechanism and rods in gear change lever bracket. On five-speed gearbox lubricate the bell crank pivot housing with gear oil (C.600).
(*Job No. A.2 in Instruction Book*)

LUBRICATION OPERATIONS ARE IN ORDINARY TYPE ;

EVERY 250 MILES—continued

15. Check gearbox oil level and top up as required with gear oil (C.600). Allow time for the oil to find its level before re-checking.

(*Job No. A.2 in Instruction Book*)

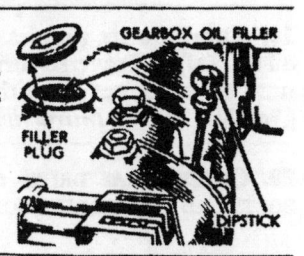

16. Lubricate two nipples on each periscope with C.600, using the pressure gun.
(*Job No. A.2 in Instruction Book*)

17. Lubricate clip and hinge pin of each periscope glass container with engine oil (10 H.D.).
(*Job No. A.2 in Instruction Book*)

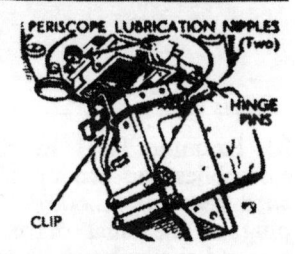

18. Lubricate gun trunnion bearings with engine oil (10 H.D.)—or with C.600 if pressure nipples are fitted.

Note that on later vehicles with front Besa gun a lubricator is fitted to the bottom trunnion of the No. 19 M.G. front Besa gimbal mounting (C.600).

(*Job No. A.2 in Instruction Book*)

Upper illustration on right shows 2-Pdr. Gun Trunnion Bearing, Churchill I and II.

Lower illustration 6-Pdr. Gun Trunnion Bearing Churchill III and IV (2 nipples—one each side.)

INSPECTIONS AND ADJUSTMENTS IN CAPITAL LETTERS

EVERY 250 MILES—continued

19. Lubricate six nipples on turret race inner ring with gear oil (C.600). Traverse turret slowly and continue to lubricate until oil oozes out beneath the race *all round*.
(*Job No. A.2 in Instruction Book*)

20. CHECK FINAL DRIVE COUPLING BOLTS AND TIGHTEN IF NECESSARY.

FINAL DRIVE COUPLING BOLTS

21. Examine level in final drive assemblies and top up as required with gear oil (C.600). The filler plug in the final drive casing is reached through a hole in the hull side plate closed by a bolted on cover. Remove the cover, and rotate the final drive until the filler plug comes opposite the hole.
(*Job No. A.2 in Instruction Book*)

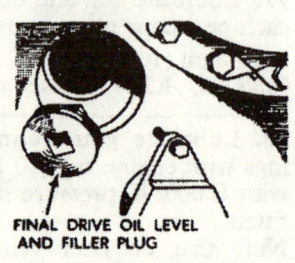

FINAL DRIVE OIL LEVEL AND FILLER PLUG

22. START AUXILIARY PETROL ELECTRIC GENERATOR AND CHECK FOR CORRECT OPERATION.

23. Lubricate auxiliary tank release mechanism at rear of hull with gear oil (C.600).
(*Job No. A.2 in Instruction Book*)

AUXILIARY TANK RELEASE MECHANISM

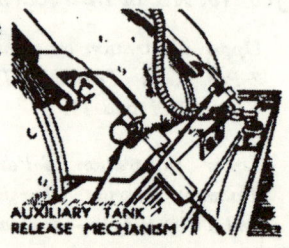

LUBRICATION OPERATIONS ARE IN ORDINARY TYPE ;

EVERY 250 MILES—continued

24. Lubricate hinges on loading doors and vision port with gear oil (C.600). Only on vehicles fitted with pressure nipples for this purpose.

(*Job No. A.2 in Instruction Book*)

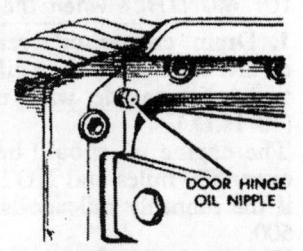

25. Lubricate all moving parts of turret traverse handle (10 H.D.).

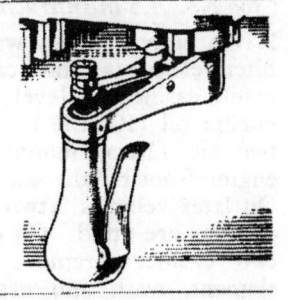

26. CHECK BOGIE ATTACHING BOLTS (88 PLACES) FOR TIGHTNESS.

(*Job No. A.17 in Instruction Book*)

27. REMOVE DRAIN PLUG FROM BASE OF ROTARY BASE JUNCTION BOX AND ALLOW ANY WATER TO DRAIN AWAY.

Position of drain plug shown in Fig. 16, page 60.

INSPECTIONS AND ADJUSTMENTS IN CAPITAL LETTERS

EVERY 500 MILES
(Or MONTHLY when the monthly mileage is less than 500)

1. Drain engine oil, renew A.C. oil filter elements, clean oil strainers, and refill engine with engine oil (10 H.D.).

The engine oil should be changed every 500 miles and NOT monthly if the monthly mileage is less than 500.

(*Job No. A.5 in Instruction Book*)

2. On early vehicles wash out air filter elements and casings and re-fill casings to level mark with engine oil (10 H.D.). There are two air filters, mounted on the engine front bulkheads.

On later vehicles "two-stage" air cleaners are fitted. On each of the two cleaners, remove the filter element, wash and drain. Wash out the oil pan and refill with engine oil (10 H.D.) to level mark, replacing empty compensator bowl when re-assembling. Examine the cork seal in the head and the neoprene seal in the shell. Tighten side wing nuts fully. Empty bottom dust container and ensure that dust ejection slots are free and that no oil drips into the centrifugal conical pan.

Note.—Do not tilt a cleaner excessively when detached and always remove the compensator when filling.

See special section—" Effects of Operating Conditions " on page 65.

(*Job No. A.3 in Instruction Book*)

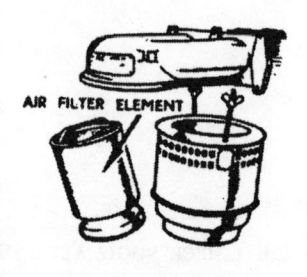

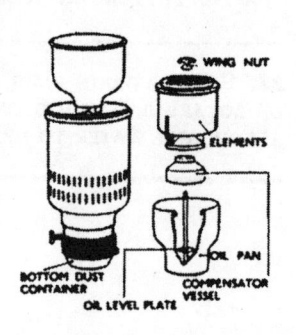

LUBRICATION OPERATIONS ARE IN ORDINARY TYPE;

EVERY 500 MILES—continued

3. CLEAN CONTACT BREAKER POINTS AND RESET IF NECESSARY. GAP .010 TO .012 IN.

(*Job No. A.7 in Instruction Book*)

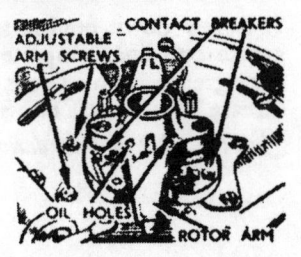

---

4. Screw down governor grease cups (2) one complete turn. If the grease cups need refilling, Grease No. 3 must be used.

(*See illustration to item 3, page 66*).

Job No. A.2 in Instruction Book)

---

5. CLEAN STRAINERS (2) IN PETROL PUMP BY REMOVING METAL FILTER BOWLS AND FLUSHING OUT. IT IS NOT NECESSARY TO REMOVE THE STRAINERS. (ACCESS THROUGH TRAP 2—SEE PAGE 60).

(*Job No. A.3 in Instruction Book*)

---

6. FLUSH OUT CARBURETTOR BOWLS AND CLEAN STRAINERS IN BANJO CONNECTIONS AT CARBURETTOR BOWLS (4).

(*Job No. A.4 in Instruction Book*)

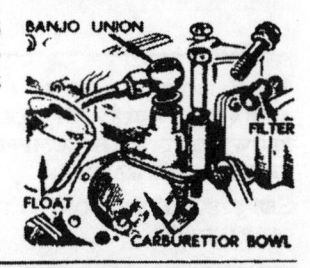

INSPECTIONS AND ADJUSTMENTS IN CAPITAL LETTERS

EVERY 500 MILES—continued

7. Screw down engine tachometer grease cup one complete turn. If the grease cup needs refilling, Grease No. 3 must be used.

(*Job No. A.2 in Instruction Book*)

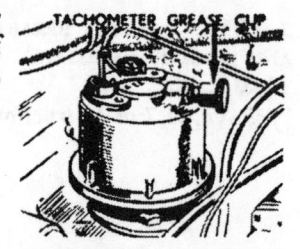

8. REMOVE PLUG FROM AIR COMPRESSOR OUTLET FILTER AND DRAIN OFF WATER.

(*Job No. A.15 in Instruction Book*)

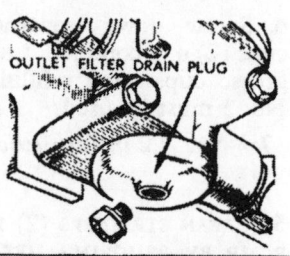

9. EXAMINE UNLOADER VALVE ON THE AIR COMPRESSOR RESERVOIR; LAP VALVE SEATS IF NECESSARY WITH METAL POLISH.

(*Job No. B.21 in Instruction Book*)

10. REMOVE AIR INLET LOUVRES AND CLEAN AIR PASSAGES IN RADIATOR CORES. IF POSSIBLE THE AIR PASSAGES SHOULD BE BLOWN OUT WITH COMPRESSED AIR FROM THE CENTRE OF THE VEHICLE OUTWARDS.

SEE SPECIAL SECTION "EFFECTS OF OPERATING CONDITIONS" ON PAGE 65.

LUBRICATION OPERATIONS ARE IN ORDINARY TYPE ;

EVERY 500 MILES—continued

11. Lubricate two nipples on steering servo cylinders with gear oil (C.600).
(*Job No. A.2 in Instruction Book*)

12. Refill oil cups (2) of steering servo cylinders with engine oil (10 H.D.).
(*Job No. A.2 in Instruction Book*)

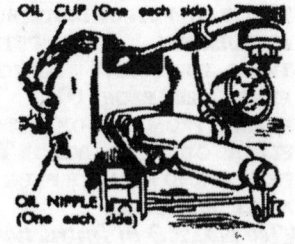

13. Lubricate pivots and ratchet release on hand-brake mechanism with engine oil (10 H.D.).
(*Job No. A.2 in Instruction Book*)

14. Drain oil from air compressor and refill with engine oil (10 H.D.). Fill to level of filler plug. (*See illustration to item 9 page 63*).
(*Job No. A.2 in Instruction Book*)

15. REMOVE CYLINDER HEAD OF AIR COMPRESSOR AND RENEW INLET AND OUTLET DISC VALVES.
THIS MUST BE DONE AFTER EVERY 500 MILES—NOT MONTHLY IF THE MILEAGE IS LESS THAN 500.
(*Job No. B.21 in Instruction Book*)

16. WASH FILTER CLOTH OF AIR COMPRESSOR INLET.
(NOTE. SEE SPECIAL SECTION "EFFECTS OF OPERATING CONDITIONS" ON PAGE 65).
(*Job No. B.21 in Instruction Book*)

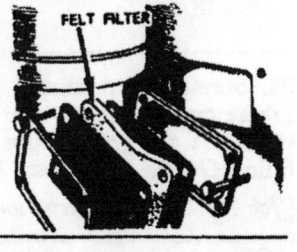

INSPECTIONS AND ADJUSTMENTS IN CAPITAL LETTERS

EVERY 500 MILES—continued

17. DRAIN PETROL DISTRIBUTION BOX AS FOLLOWS. TURN PETROL CONTROLS TO "OFF." REMOVE DRAIN PLUG (SEE PAGE 60). TURN ON PETROL FROM ONE OF THE TANKS TO FLUSH OUT THE BOX. TURN OFF PETROL AND REPLACE PLUG SECURELY AFTER FLUSHING.
(*Job No. A.3 in Instruction Book*)

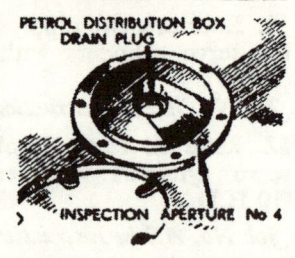

18. EXAMINE TIGHTNESS AND CLEANLINESS OF TERMINALS IN THE ROTARY BASE JUNCTION BOX.

19. Screw down main dynamo grease cups (if fitted) one complete turn. If the grease cups need refilling, Grease No. 3 must be used. (*See illustration to item 5, page 66*).
(*Job No. A.2 in Instruction Book*)

20. CLEAN DISTRIBUTOR COVERS AND CHECK WIRING CONNECTIONS FOR TIGHTNESS.

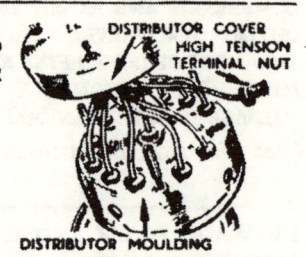

21. Screw down turret generator grease cups (if fitted) one complete turn. If the grease cups need refilling *Grease No. 3 must be used*. (*See illustration to item 11, page 68*).
(*Job No. A.2 in Instruction Book*)

LUBRICATION OPERATIONS ARE IN ORDINARY TYPE;

76

EVERY 500 MILES—continued

22. Fill oil cup on starter motor with engine oil (10 H.D.). Access through Trap 2 (see page 60).

(*Job No. A.2 in Instruction Book*)

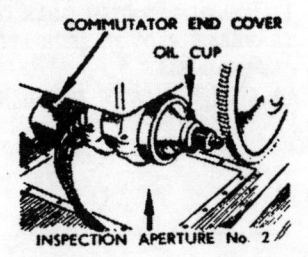

INSPECTIONS AND ADJUSTMENTS IN CAPITAL LETTERS

EVERY 1,000 MILES

1. EXAMINE BRUSH GEAR OF TURRET TRAVERSE MOTOR FOR WEAR AND CLEANLINESS.
(*Job No. A.16 in Instruction Book*)

2. EXAMINE BRUSHES OF STARTER MOTOR, MAIN DYNAMO AND TURRET GENERATOR FOR WEAR AND CLEANLINESS. (ACCESS TO STARTER MOTOR THROUGH TRAP 2—SEE PAGE 60).
(*Job No. A.16 in Instruction Book*)

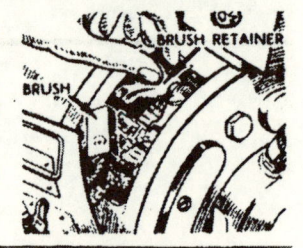

3. REMOVE FILTER DISCS FROM COMPRESSOR OUTLET FILTER, CLEAN THOROUGHLY IN PETROL AND REPLACE.
(*Job No. A.15 in Instruction Book*)

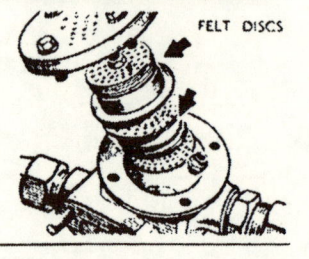

4. REMOVE, CLEAN AND RE-SET SPARKING PLUGS (GAP—.018 TO .020 IN.).
(*Job No. A.9 in Instruction Book*)

5. Drain sump of engine of auxiliary petrol electric generator and refill with engine oil (10 H.D.) to bottom thread of filler plug.
(*Job No. A.2 in Instruction Book*)

(*See illustration to item 13, page 64*).

LUBRICATION OPERATIONS ARE IN ORDINARY TYPE;

EVERY 1,000 MILES—continued

6. RENEW FLEXIBLE DRIVE SHAFT TO AMAL PUMP WHERE FITTED. TAKE GREAT CARE TO LOCATE THE CLIPS CORRECTLY SO THAT SHARP BENDS ARE AVOIDED. (ACCESS THROUGH TRAP 2—SEE PAGE 60).

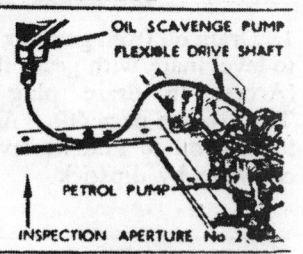

7. Lubricate idler wheels with gear oil (C.600) until oil exudes from the bearing. A nipple is provided on each idler for the pressure gun.

(*Job No. A.2 in Instruction Book*)

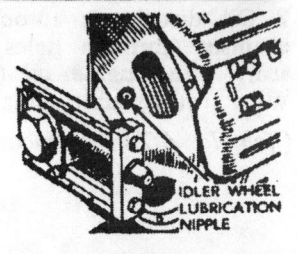

INSPECTIONS AND ADJUSTMENTS IN CAPITAL LETTERS

EVERY 2,000 MILES

1. Drain oil from gearbox and refill to level mark with gear oil (C.600). (Access to drain plug through Trap 1—see page 60). Allow time for the oil to find its level before checking by dipstick.

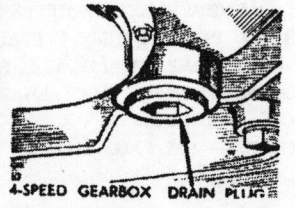

4-SPEED GEARBOX DRAIN PLUG

2. Oil distributor automatic mechanism (through holes in rotor arm) with engine oil (10 H.D.). Wipe arm dry after oiling.

(See illustration to item 3, page 73).

(Job No. A.2 in Instruction Book)

GENERAL HINTS AND TIPS

Don't drive fast on frozen, bumpy ground, or you may find the vehicle getting out of control.

* * *

Don't run with slack tracks if it can possibly be avoided. Slack tracks affect steering.

* * *

Watch the engine rev. counter when driving—and make sure the reading doesn't exceed 2,000 r.p.m. when going downhill.

* * *

If the CO_2 fire extinguisher bottles are used, obtain a replacement *immediately*.

* * *

The air servo motors are provided to make clutch and steering control easier. If the air pressure is lost, the vehicle can still be handled effectively without them although the controls will not be so light.

* * *

Don't attempt obstacle crossing or severe cross-country work with the petrol at a low level. Some of it is trapped when the vehicle is tilted, leaving the suction pipes high and dry.

* * *

Don't run with the cooling system short of water. Check frequently on *both* sides of the vehicle. A dry engine may mean cracked castings and blown gaskets.

* * *

Don't "ride" the clutch. The throw-out gear is air-assisted, and the air valve may open if the pedal is used as a foot-rest. Result—slipping and overheating.

* * *

Use a funnel for oil-filling. Spilt oil has often been mistaken for oil pipe leakage.

* * *

Open the water filler caps slowly after a run. The system operates under pressure, and a jet of steam or boiling water doesn't make a pleasant shower.

After checking and topping-up the oil level (especially after a long stand), run the engine for a few moments and then check the level again.

* * *

Handle the escape hatches carefully, and make sure they are securely fastened when open. If they drop on someone, they hurt.

* * *

Endeavour to keep the petrol economy light " out " when driving. This saves petrol.

* * *

Never turn the steering bar when the engine is running unless you desire to turn the vehicle.

* * *

Do not leave the vehicle parked, especially on a slope, with the hand-brake on only. Always engage a low gear.

* * *

Always close and lock the engine compartment hatches before moving off. This is necessary for correct engine cooling.

* * *

An accumulation of water or other fluids in the engine compartment can be cleared by opening the petrol dump valve. The control is in the fighting compartment.

* * *

The correct place for the tools is in the tool-box behind the driver's seat. More tools are lost by being left about the vehicle than are " scrounged " by other people.

* * *

Keep your tool equipment in good order. If a tool is damaged, report it and get a replacement. An inefficient tool is useless—and is not a good excuse for bad work.

* * *

When crossing obstacles meet them at right-angles whenever possible—this minimises side stress on the tracks.

* * *

Always remember that the Churchill is a heavy tank, and start from rest smoothly. A violent start imposes undue strains throughout the transmission.

Before starting, make sure that the two temperature indicators are recording. Both should indicate approximately 185° F. Remember, too, to keep an occasional eye on them while running. They will give warning of any fault developing in the cooling system.

* * *

Do not completely empty any petrol tank before changing over to another supply. The heavy suction caused when a tank is dry will overload the petrol pump drive. (This applies only to vehicles fitted with the *flexibly* driven Amal pump.)

* * *

Check the clutch pedal clearance daily. If the "free travel" is not correct the clutch will slip. The "free travel" will vary very little if the clutch is properly used.

* * *

Always use "Lockheed Racing Green Fluid" or Hydraulic Brake Fluid No. 4 for topping-up the Lockheed systems. There is no permissible substitute.

* * *

When filling the water-systems or washing down the vehicle, try to prevent water accumulating in the engine compartment. This will be blown into the steering brakes by the fan, and the brakes will be ineffective until they are dried out.

RAIL TRANSPORT
Before transporting the Churchill Tank by rail the overall width must be reduced by removing the spare track plates, the tow rope shackle pins (unless they are welded into position) and the air inlet louvres. Details are given in the Instruction Book.

LUBRICATION CHART

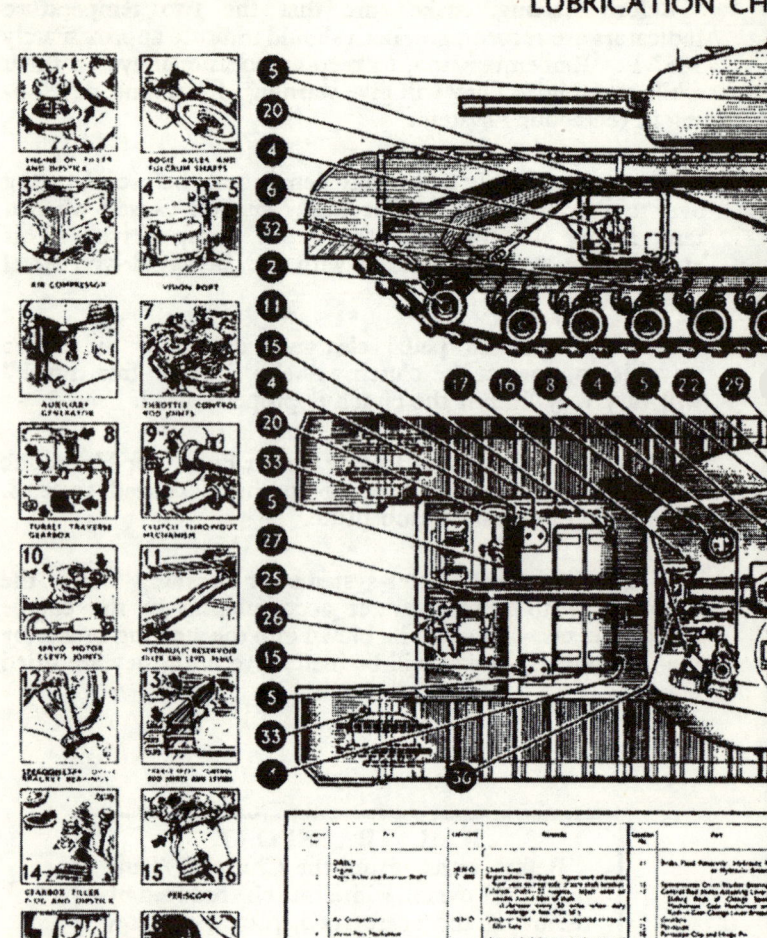

CHURCHILL I, II, III & IV

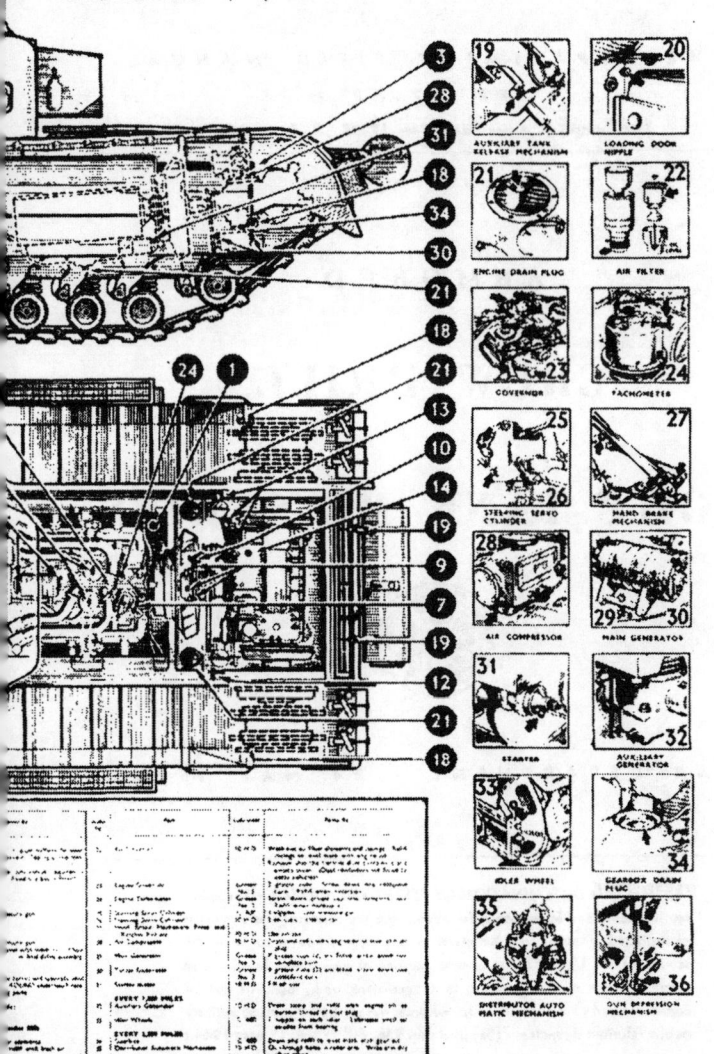

WAR DEPARTMENT FIELD MANUAL
FM 17-68
This manual supersedes FM 17-68, 8 June 1943

ARMORED

CREW DRILL

LIGHT TANK M5 SERIES

WAR DEPARTMENT · 24 MAY 1944

RESTRICTED *DISSEMINATION OF RESTRICTED MATTER.* —The Information contained in restricted documents and the essential characteristics of restricted material may be given to any person known to be in the service of the United States and to persons of undoubted loyalty and discretion who are cooperating in Government work, but will not be communicated to the public or to the press except by authorized military public relations agencies. (See also par. 23b, AR 380–5, 15 Mar 1944.)

United States Government Printing Office
Washington 1944

WAR DEPARTMENT,
Washington 25, D. C., 24 May 1944.

FM 17-68, Armored Field Manual, Crew Drill, Light Tank M5 Series, is published for the information and guidance of all concerned.

[A. G. 300.7 (24 May 44).]

By order of the Secretary of War:

G. C. MARSHALL,
Chief of Staff.

Official:

J. A. ULIO,
Major General.
The Adjutant General.

Distribution:

As prescribed in paragraph 9a, FM 21-6, except Armd Sch (500); D 2, 7, 17 (10); R, 17 (10); Bn 2, 17 (5); IC 17 (5), (20).

IC 17: T/O & E, Hq Co, Armd Div; 17-20-1, Hq and Hq Co, Combat Comd, Armd Div (5); 17-17, Light Tk Co, Tank Bn and Light Tank Co, Cav Mecz Sq (20).

(For explanation of symbols, see FM 21-6.)

TABLE OF CONTENTS

		Paragraph	Page
Section	I. General	1–2	1
	II. Crew Composition and Formation	3–4	4
	III. Crew Control	5–6	6
	IV. Crew Drill	7–12	10
	V. Service of the Piece	13–18	19
	VI. Mounted Action	19–23	31
	VII. Dismounted Action	24–29	37
	VIII. Evacuation of Casualties from Tanks	30–33	46
	IX. Inspections and Maintenance	34–40	51
	X. Sight Adjustment	41–42	68
	XI. Destruction of Materiel and Equipment	43–50	71

RESTRICTED

FM 17-68

WAR DEPARTMENT FIELD MANUAL

CREW DRILL
LIGHT TANK M5 SERIES*

This manual supersedes FM 17-68, 8 June 1943.

Section I

GENERAL

---

1. PURPOSE AND SCOPE. This manual is designed to present instructional material for the platoon leader and tank commander in training members of the crew of the light tanks M5 and M5A1 for combat. It is to be used as a guide to achieve orderly, disciplined, efficient execution of mounted and dismounted action, and precision, accuracy, and speed in service of the piece. It provides a logical and thorough routine for all inspections of the vehicle and its equipment.

2. REFERENCES. See appendix.

---

* "For definition of military terms not defined in this manual see TM 20-205".

Figure 1. Light tank M5A1, front and rear views.

2

Figure 2. Light tank M5A1.

Section II

COMPOSITION AND FORMATIONS

3. COMPOSITION. The light tank crew is composed of four members:

Tank commander (loader or assistant gunner and tends voice radio)	LIEUTENANT or SERGEANT
Gunner	GUNNER
Bow Gunner (assistant driver, and radio operator in tanks equipped with SCR-506)	BOG
Driver	DRIVER

Figure 3. Dismounted posts, light tank crew.

4. FORMATIONS. a. Dismounted posts. The crew forms at attention in one rank. (See figure 3). The tank commander takes his post 2 yards in front of the right track, facing the front. The bow gunner, gunner and driver in order, take positions at the left of the tank commander at close interval.

b. Mounted posts. The crew forms mounted as follows:

(1) *Tank commander.* In the turret, standing on the floor, sitting or standing on the seat.

(2) *Bow gunner.* In the bow gunner's seat.

(3) *Gunner.* In the turret, seated at the left of the 37-mm gun with his left hand on the power traverse control handle, his right hand on the elevating handwheel, and his head against the headrest of the periscope.

(4) *Driver.* In the driver's seat.

Section III

CREW CONTROL

5. OPERATION OF INTERPHONE AND RADIO. The crew must practice continually with the interphone to attain its maximum value during combat. It is available for use at any time during the operation of the tank, but its use interrupts radio communication. Due to the differences in location of the radio and interphone facilities of the light tank M5 and M5A1, slightly different procedures are necessary in crew control. For tank signals see FM 17-5.

a. M5 light tank. As standing operating procedure, after mounting, the radio is turned on by either the driver or gunner without command, and the headsets and microphones are tested. See Preliminary Inspection for Radio Sets SCR-508, SCR-528, and SCR-538.

(1) Each crew member (except the gunner who has no interphone connection) inserts the plugs of the short cords extending from his headset into the break-away plug of the headset extension cord of his interphone control box. The microphone (of either type, throat or lip) is adjusted in place to produce maximum clarity, and connected to the break-away plug on the microphone cord of the control box.

(2) The tank commander turns his **RADIO INTERPHONE** switch to **INTERPHONE**, depresses the switch on his microphone cord, and orders: CHECK INTERPHONE. NOTE: This command is used when the crew mounts by any other method than the formalized drills given in paragraphs 8 and 25. In those drills the "Ready" report constitutes the interphone check.

(3) The bow gunner and driver, in turn, depress their microphone switches and report: BOG CHECK, DRIVER CHECK. During this procedure, each adjusts

101

the volume control on his interphone control box to the desired level. *Care is taken that the microphone switch does not remain in locked position thus burning out the dynamotor.* In stowing, check to see that the suspension strap is not wrapped around the hand switch, pressing down the switch button.

(4) Interphone control box positions are located as follows:

(a) *Driver.* On wall of hull to his left.

(b) *Bow gunner.* On wall of hull to his right.

(c) *Tank commander.* On rear of turret roof between the hatches. He controls his transmission by manipulating the switch on the control box marked RADIO-INTERPHONE to the type of transmission desired.

(5) The RADIO-INTERPHONE switches on the transmitter and on all control boxes, except the tank commander's control *box, are set on RADIO. This is the normal position for interphone operation.* The tank commander's switch is set on RADIO most of the time, he will change it to INTERPHONE only as he desires to communicate with his crew. Except in an emergency, *no one but the tank commander* may operate the interphone switch on his control box. In an emergency, a member of the tank crew may break in while the tank commander is on RADIO by turning his control box switch to INTERPHONE. When the tank commander is on RADIO, turning any RADIO-INTERPHONE switch to INTERPHONE will interrupt the commander's radio communication and establish interphone communication.

(6) It is the duty of each man to check his personal interphone equipment upon mounting the tank, see that it is properly maintained, and report any difficulties to the tank commander. Definite tank control commands and terminology are set forth in the following paragraph. The desirability and necessity of adhering to this specific language cannot be overemphasized. General conversation on interphones causes misunderstanding and disorder and is harmful to discipline.

b. M5A1 light tank. In this and later models, the radio is located in the turret bulge, immediately accessible to the tank commander. The gunner's interphone control box is on the roof of the turret in front of him. The tank commander's control box is on the right side of the turret. On mounting the M5A1 tank, the crew follows the same procedure as that prescribed for the M5 model with the following exceptions: the tank commander turns on the receiver, transmitter or amplifier, and sets the channel number button. Upon the command CHECK INTERPHONE, the gunner reports first, followed in order by the bow gunner and driver.

NOTE: In tanks equipped with SCR-506, two members of the crew (Bog and gunner) are qualified radio operators. The principal duty of the Bog is to operate this set, and interphone procedure is modified as required to enable him to perform his duties.

6. INTERPHONE LANGUAGE. a. Terms.

Tank Commander	LIEUTENANT or SERGEANT
Driver	DRIVER
Gunner	GUNNER
Bow Gunner	BOG
Any tank	TANK
Armored car	ARMORED CAR
Any unarmored vehicle	TRUCK
Infantry	DOUGHS
Machine gun	MACHINE GUN
Airplane	PLANE

b. Commands for movement of tank.

To move the tank forward	DRIVER MOVE OUT
To halt the tank	DRIVER STOP
To reverse the tank	DRIVER REVERSE
To decrease speed	DRIVER SLOW DOWN
To turn right 90°	DRIVER CLOCK 3 --- STEADY-Y-Y-Y . . . ON
To turn left 60°	DRIVER CLOCK 10--- STEADY-Y-Y-Y . . . ON

To turn right (left) 180°	DRIVER CLOCK 6 RIGHT (LEFT)---STEADY-Y-Y-Y ... ON
To have driver move toward a terrain feature or reference point. The tank being headed in proper direction	DRIVER MARCH ON WHITE HOUSE (HILL, DEAD TREE, ETC.)
To follow in column	DRIVER FOLLOW THAT TANK (DRIVER FOLLOW TANK NO B-9)
To follow on road or trail	DRIVER RIGHT ON ROAD (DRIVER RIGHT ON TRAIL)
To start engines	DRIVER CRANK UP
To stop the engine	DRIVER CUT ENGINE
To proceed at same speed	DRIVER STEADY

  c. **Comands for control of turret.**

To traverse the turret	GUNNER TRAVERSE LEFT (RIGHT)
To stop turret traverse	GUNNER STEADY-Y-Y-Y ... ON

  d. **Fire orders.** See FM 17-12 .

Section IV

CREW DRILL

---

7. DISMOUNTED DRILL. a. To form light tank crew. Being dismounted, the crew takes dismounted posts (figure 3) at the command FALL IN.

b. To break ranks. At the command FALL OUT, the crew breaks ranks. Crew members habitually fall out to the right of the tank.

c. To call off. At the command CALL OFF, the members of the crew call off in turn as follows:

(1) Tank Commander _____ "SERGEANT" (or "LIEUTENANT")
(2) Bow Gunner _____ "BOG"
(3) Gunner _____ "GUNNER"
(4) Driver _____ "DRIVER"

d. To change designations and duties. (1) At the command FALL OUT SERGEANT (BOG) (GUNNER) —

(*a*) The man designated to fall out moves by the rear to the left flank position and becomes driver.

(*b*) The crew members on the left of the vacated post move smartly to the right one position, ready to call off their new designations.

(*c*) The acting tank commander starts calling off as soon as the crew is re-formed in line.

(2) The movement is executed by having any member of the crew fall out except the driver.

(3) All movements are executed with snap and precision and at double time.

8. TO MOUNT LIGHT TANK CREW. Crew being at dismounted posts (figure 4).

Tank Commander	Gunner	Bow Gunner	Driver
Command PREPARE to MOUNT.			
About face.	About face.	About face.	About face.
Command MOUNT.			
Mount to right fender.	Mount to left fender.	Stand fast.	Stand fast.
Mount to right sponson.	Mount to left sponson.	Mount to right fender.	Mount to left fender.
Enter turret and take mounted post.	Enter turret and take mounted post. Turn on radio.	Enter Bog's hatch and take mounted post.	Enter Driver's hatch and take mounted post. Turn on master switch.
Connect breakaway plugs.	Connect breakaway plugs.	Connect breakaway plugs.	Connect breakaway plugs.
Command REPORT.	Report "Gunner Ready".	Report "Bog Ready".	Report "Driver Ready".

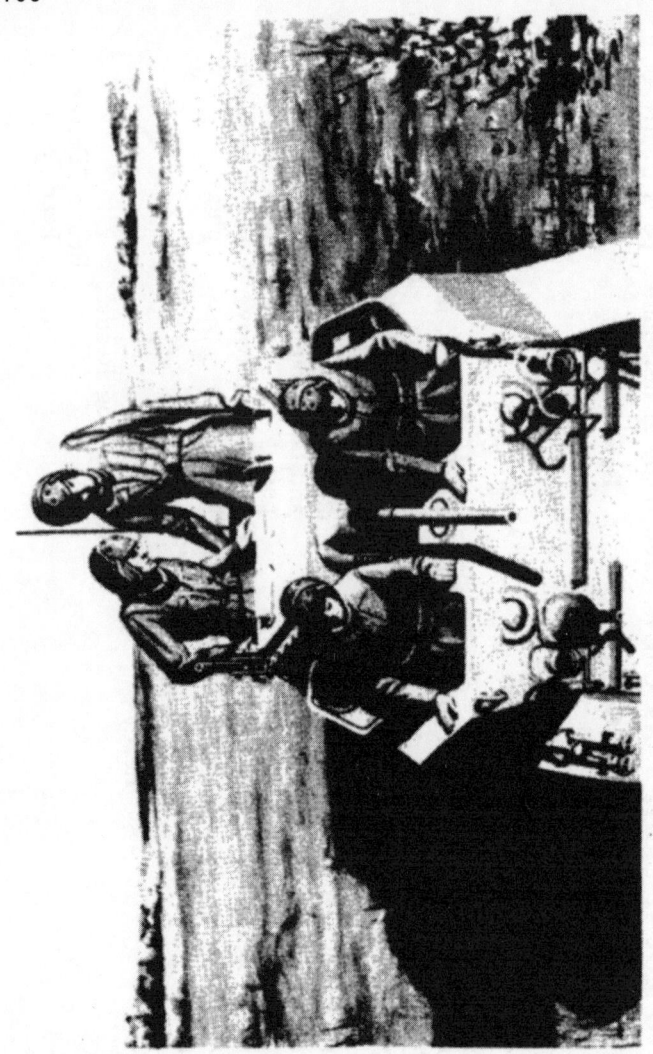

Figure 4. Mounting through hatches, light tank M5A1.

9. TO CLOSE AND OPEN HATCHES.

a. Crew being at mounted posts. TO CLOSE HATCHES.

Tank Commander	Gunner	Bow Gunner	Driver
Check that turret is in straight ahead position; order Gunner to traverse as necessary. Command CLOSE HATCHES.	Traverse turret as ordered by Tank Commander.		
Close hatch. Raise periscopes. Command REPORT.	Close hatch. Raise periscope. Report "Gunner Ready".	Close hatch. Raise periscope.	Close hatch. Raise periscope.
		Report "Bog Ready".	Report "Driver Ready".

b. Crew being mounted and hatches closed, TO OPEN HATCHES —

Tank Commander	Gunner	Bow Gunner	Driver
Check that turret is in straight ahead position; order Gunner to traverse as necessary.	Traverse gun as ordered by Tank Commander.		

13

Tank Commander	Gunner	Bow Gunner	Driver
Command OPEN HATCHES. Lower periscopes. Open hatch. Command REPORT.	Open hatch. Report "Gunner Ready".	Lower periscope. Open hatch. Report "Bog Ready".	Lower periscope. Open hatch. Report "Driver Ready".

10. TO DISMOUNT TANK CREW. Hatches being open, to dismount without vehicular weapons—

Tank Commander	Gunner	Bow Gunner	Driver
Command PREPARE TO DISMOUNT	Turn off radio.		Turn off master switch.
Disconnect breakaway plugs. Command DISMOUNT.	Disconnect breakaway plugs.	Disconnect breakaway plugs.	Disconnect breakaway plugs.

Tank Commander	Gunner	Bow Gunner	Driver
Dismount to right sponson. Dismount to right fender. Take dismounted post.	Dismount to left sponson. Dismount to left fender. Take dismounted post.	Dismount to right fender. Take dismounted post.	Dismount to left fender. Take dismounted post.

11. TO DISMOUNT THROUGH ESCAPE HATCH. (Applicable only to M5A1 and later models equipped with escape hatch.) Crew being at mounted posts, to dismount without vehicular weapons.

Tank Commander	Gunner	Bow Gunner	Driver
Command THROUGH ESCAPE HATCH, PREPARE TO DISMOUNT.	Turn off radio.		Turn off master switch.
Disconnect breakaway plugs.	Disconnect breakaway plugs. Raise recoil guard. Elevate breech of gun.	Disconnect breakaway plugs. Unclamp spare parts box from escape hatch door and open the door.	Disconnect breakaway plugs.

15

Tank Commander	Gunner	Bow Gunner	Driver
Command DISMOUNT.	Move to right side of turret.		
Enter Bog's compartment, feet first. Dismount through escape hatch.		Dismount through escape hatch.	
	Enter Bog's compartment feet first. Dismount through escape hatch.	Crawl from under tank; take dismounted post.	
Crawl from under tank; take dismounted post.			Enter Bog's compartment. Dismount through escape hatch.
	Crawl from under tank; take dismounted post.		Crawl from under tank; take dismounted post.

NOTE: In some models of the light tank the Driver cannot cross the transmission into the Bog's compartment. In such cases he dismounts by passing through the turret compartment. (The Gunner must move an ammunition box from the turret opening before this can be done.)

12. PEP DRILL. To maintain the interest of crew members, frequent and unexpected periods of pep drill are interspersed in the crew drill and simulated firing routines. Pep drill is a series of precision movements executed at high speed and terminating at the position of attention, either mounted or dismounted. For example, the crews being dismounted, the platoon commander may suddenly command, IN FRONT OF YOUR TANKS, FALL IN; MOUNT: DISMOUNT; ON THE LEFT OF YOUR TANKS, FALL IN; FORWARD, MARCH; BY THE LEFT FLANK, MARCH; TO THE REAR, MARCH; MOUNT! (figure 5). Preparatory commands for mounting and dismounting are eliminated in this drill. The posts of all crew members are changed frequently. Pep drill freshens the interest of the crews, trains them to be agile in and around the tank, and increases their coordination and physical development.

(a) DISMOUNT; on the left of your tanks, FALL IN.

(b) MOUNT.

Figure 5. Pep drill.

Section V

SERVICE OF THE PIECE

---

13. GUN CREW. a. The gun crew, 37-mm tank gun, consists of the gunner who aims and fires the piece and the assistant gunner (tank commander) who loads the piece when not observing, and controls and adjusts fire.

b. Training in service of the piece must stress rapidity and precision of movement and teamwork.

14. POSITIONS OF GUN CREW. Positions of the gun crew are as prescribed in paragraph 4b.

15. OPERATION OF GUN. a. To open the breech. Grasp the T-handle of the crank under the breechblock and pull down until the breechblock is locked in the open position.

b. To close the breech. The insertion of a round in the gun trips the extractors and causes the breechblock to close automatically. For this reason use care in closing the breech manually as follows:

(1) Insert an empty cartridge case in the breech, base foremost, and trip the extractors. The breechblock will close, pushing the cartridge case upward.

(2) If an empty cartridge case is not available a block of wood of the proper size may be used.

(3) If necessary, the extractors may be tripped by using the fingers. This is done by pulling down on the T-handle and overcoming the tension of the closing spring. Then push the extractor lips forward with the fingers of the free hand and allow the breechblock to rise slowly. *Utmost caution must be exercised if this method is used.*

Figure 6. Turret and gun controls on combination gun mount M44, in light tank M5A1.

Labels (left side):
- PARALLEL SIGHT LINKAGE
- SIGHTING PERISCOPE M4
- PERISCOPE HEAD REST
- STABILIZER CONTROL BOX
- TELESCOPIC SIGHT M54
- FIRING MECHANISM BRACKET ASSEMBLY
- 37-MM SOLENOID
- 37-MM GUN FIRING SWITCH
- .30 CALIBER MG FIRING SWITCH
- POWER TRAVERSE CONTROL HANDLE
- TURRET FRICTION CLAMP
- ELEVATING HANDWHEEL KNOB

Labels (right side):
- TRAVELING LOCK
- AUXILIARY POWER TRAVERSE CONTROL HANDLE
- OBSERVATION PERISCOPE
- GYRO CONTROL UNIT
- COAXIAL MACHINE GUN
- BREECH RING
- .30 CALIBER MG SOLENOID
- BREECH OPERATING MECHANISM
- BREECH (RECOIL) GUARD
- TURRET LOCK

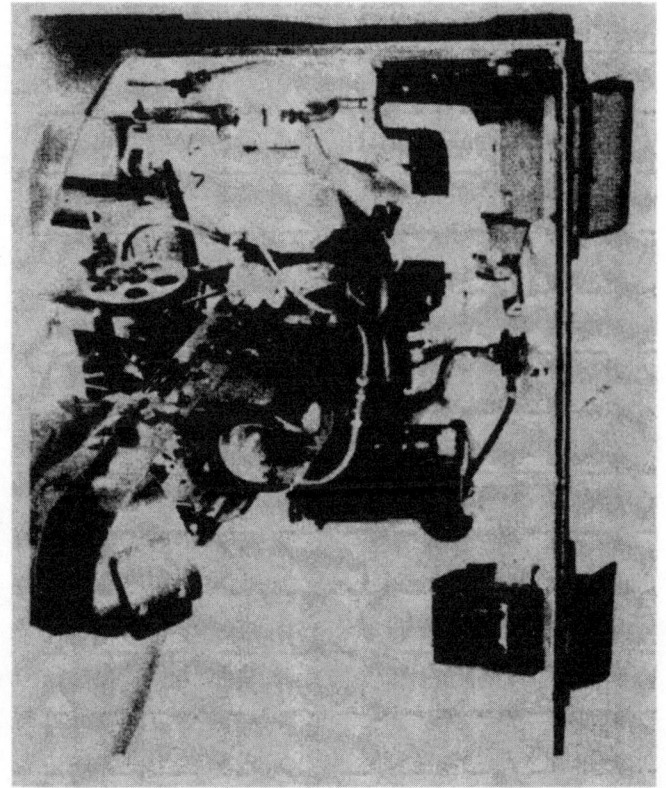

Figure 7. Left rear view of turret trainer for light tank gun crews.

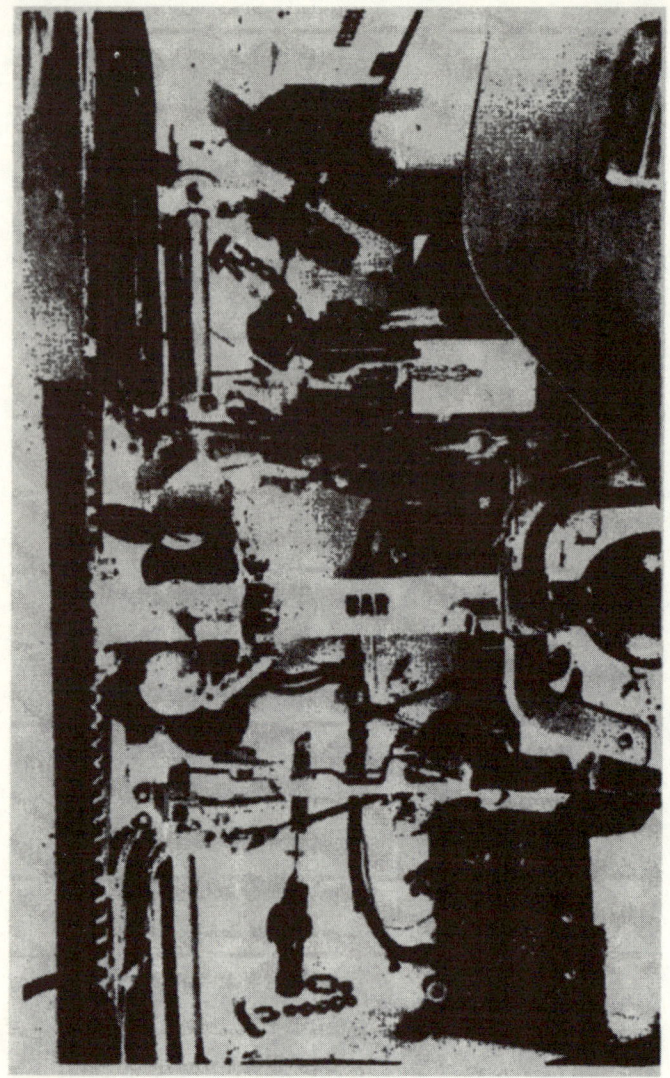

Figure 8. Driver's compartment, light tank M5.

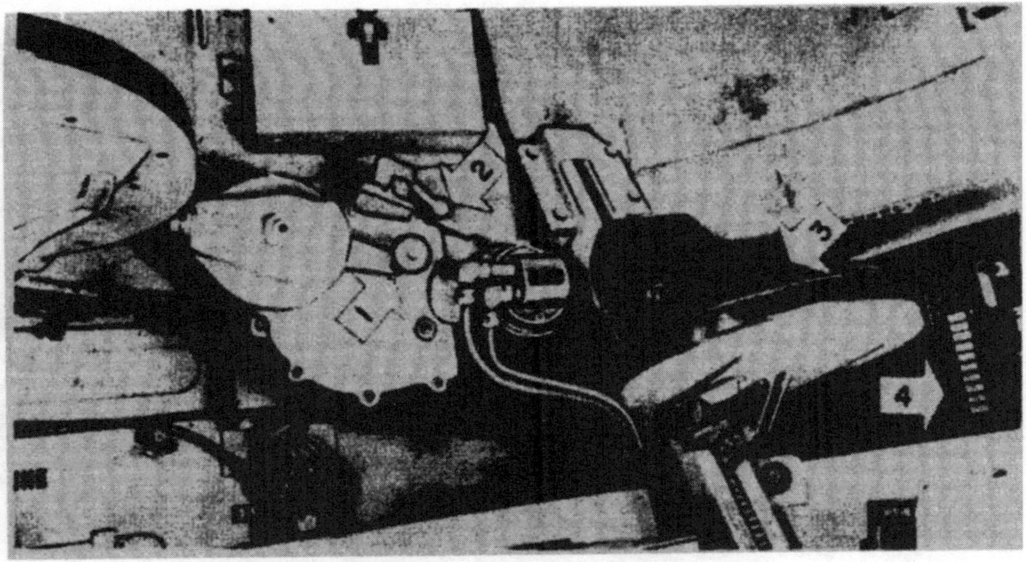

Figure 9. Turret basket, left side, light tank M5.

1. Traverse hand crank
2. Power traverse clutch lever
3. Gunner's seat
4. Radio

23

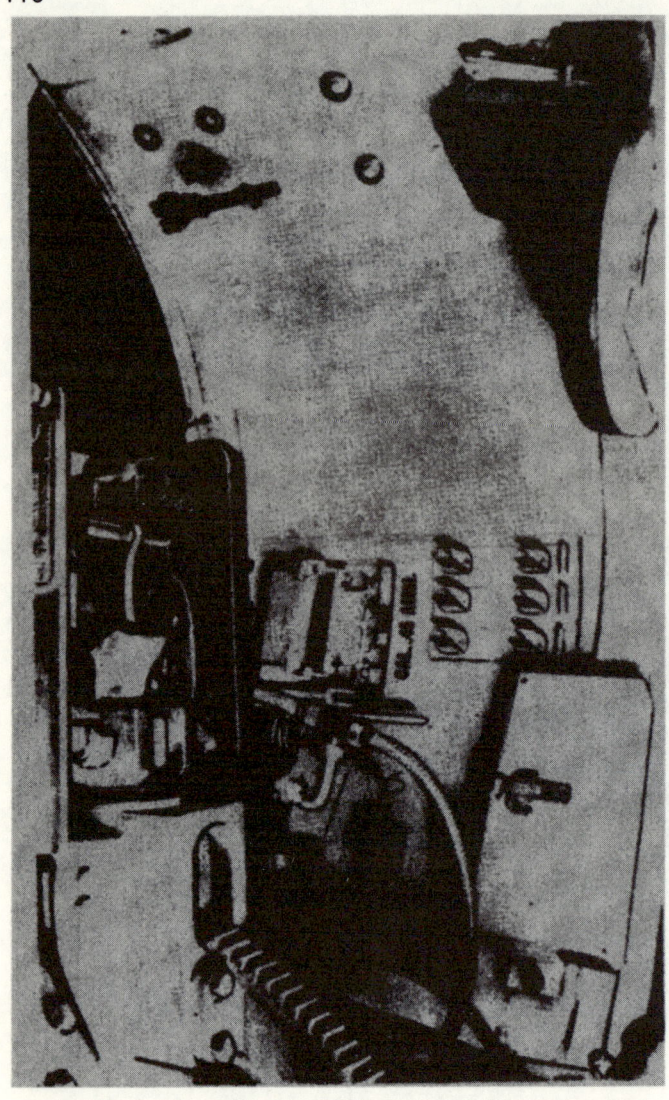

Figure 10. Turret basket, right side, tank commander's seat, light tank M5.

c. **To load.** (1) Open breech by pulling down on T-handle until breechblock is locked in the open position.

(2) Grasp projectile at the base with left hand, turn nose of projectile to the front, swing arm toward the breech and insert round into recess of breech ring. After nose of projectile is aligned in chamber, shove round forward sharply with the thumb on the base, following through with the arm moving upward to clear the closing breechlock. Keep arm above gun to avoid the path of recoil, and tap gunner on the back.

d. **To lay the gun.** Locate target through the periscope. Turn the power traverse (pistol grip) control handle, or traverse with hand crank (figure 11) in the proper direction until the vertical line of range dots (or dashes) is on the target, or the proper lead is taken. Make the final traversing motion against the greatest resistance, such as might be caused by cant in the tank. Then elevate gun until the proper range marking is on the target (see FM 17-12).

e. **To fire the gun.** Squeeze the safety trigger on the power traverse control handle and press the 37-mm gun switch simultaneously (figures 5 and 6). If the round fails to fire, proceed as in paragraphs 17 and 18. The gun can be fired manually with the firing button in the hub of the elevating handwheel.

f. **To unload an unfired round.** To unload an unfired round, pull down the T-handle and the round will be ejected into the spent case bag. The round is then returned to the rack.

g. **To unload a stuck round.** Whenever possible, rounds are removed from guns by shooting them out. When an unfired round is stuck in the gun and cannot be dislodged from the breech by using the rim of an empty cartridge case to pry it out, the gunner, under direct supervision of an officer, or the tank commander, if no officer is present, inserts the rammer in the muzzle of the gun, pushes it through the bore until it meets the round, and then shoves the round out of the gun. DO

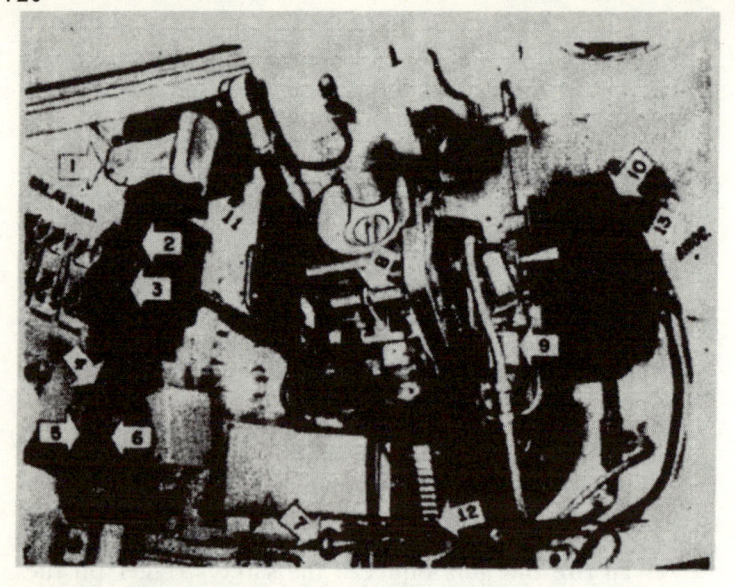

1. Periscope headrest
2. Recoil adjuster knob
3. Stiffness adjuster knob
4. Firing switches for 37-mm gun and co-axial machine gun
5. Power traverse (pistol grip) control handle
6. Safety trigger
7. Elevating handwheel
8. T-handle
9. Machine gun solenoid
10. Gyro control unit.
11. Periscopic sight M4, w/telescope M40
12. Worm and sector gear
13. Oil reservoir (filler gauge)

Figure 11. Gunner's controls, combination gun mount M23.

121

NOT STRIKE THE ROUND WITH THE RAMMER. If pressure will not dislodge a round, tap the end of the rammer staff lightly with a block of wood, alternating tapping and pressure.

h. For further malfunctions and their remedies see paragraph 17.

16. SAFETY PRECAUTIONS. a. Before firing, and during lulls in firing, inspect the gun to see that there are no obstructions in the bore.

b. In loading the gun, take care not to dent or burr the projectile by striking it against the breech ring. *Clean ammunition before loading.* Never attempt to disassemble a round.

c. The gunner fires the gun only when the loader (tank commander) has tapped him on the back to signal that the gun is loaded and ready.

d. The 37-mm gun is fired only when the tank engine(s) are running or when at least one of the turret hatches is open to carry off the powder fumes.

e. Any individual who observes a condition which makes firing dangerous will immediately call or signal "CEASE FIRING".

f. Firing will cease immediately at the command or signal "CEASE FIRING", regardless of the source of the command.

g. Before loading, the solenoid firing devices are checked to insure that they are not stuck.

h. Tank weapons, except AA gun, are fired only when the Driver's and Bog's hatches are closed.

17. MALFUNCTIONS AND REMEDIES. Malfunctions of the gun are divided into three general classes: failure to fire, failure to load, and failure to extract. The most likely malfunctions are failures to fire. Malfunctions which are not the result of broken or worn parts are generally due to carelessness and improper cleaning of the parts and ammunition.

a. Failure of gun to fire.

Cause	Remedy
1. Safety lever on "Safe".	1. Put lever on "Fire".
2. Gun out of battery.	2. Push gun into battery; paragraph 18.
3. Defective firing mechanism.	3. Repair or replace.
4. Defective ammunition.	4. Use a new round.
5. Broken or defective firing pin.	5. Replace.
6. Broken cocking lever.	6. Replace.
7. Broken cocking fork.	7. Replace.
8. Broken cocking lugs on the percussion mechanism.	8. Replace.
9. Weak firing spring.	9. Replace.

b. Failure of gun to load.

Cause	Remedy
1. Dirty round.	1. Remove round and clean.
2. Dirty chamber.	2. Remove round and clean chamber.
3. Bulged round.	3. Use a new round.
4. Dirty breechblock recess.	4. Clean.
5. Worn or broken extractor lips.	5. Replace.
6. Bent or undersize rim of round.	6. Use a new round.
7. Defective closing mechanism.	7. Repair or replace.
8. Burrs on bearing surfaces of breechblock and breech ring.	8. Report to Ordnance.
9. Gun out of battery.	9. Put gun in battery, see paragraph 18.
10. Weak closing spring.	10. Replace.

c. **Failure of gun to extract.**

Cause	Remedy
1. Broken extractor lips.	1. Pry or ram out empty case and replace extractors.
2. Undersize or bent rim of round.	2. Pry or ram out.
3. Broken operating crank.	3. Pry or ram out empty case. Replace crank.
4. Broken operating lug.	4. Pry or ram out empty round. Replace operating crank.

18. IMMEDIATE ACTION. The sequence following (trouble shooting) is done immediately after the gun fails to fire during combat, and repair or replacement of parts cannot be effected immediately, as can be done in garrison.

a. **Gun out of battery.** (1) Push gun into battery by hand, relay, and fire.

(2) If gun still fails to fire, recock by hand (use board if necessary) and attempt to fire two times. If it does not fire, remove round to determine the cause of misfire. (See AR 750-10)

(3) If gun will not go into battery by hand, fully depress muzzle. Slowly unscrew rear recoil drain plug, being careful not to remove it entirely. An excessive amount of oil might escape if the gun suddenly started back into battery. Allow oil to drain from around the plug until it will no longer flow from cylinder, then tighten filler plug and push gun back into battery by hand.

(4) If gun will not go into battery after checking recoil cylinder, check for dirt, burrs, and lack of lubrication between the bearing surfaces of the recoil cylinder slides (rails) and sleigh guides. Clean, remove burrs, lubricate, reload, relay, and fire. (For a thorough cleaning and inspection of these parts, unscrew

coupler key nut and remove coupler key; then pull barrel assembly to rear.) NOTE: It may be possible to continue gun in action by pushing it into battery each time. However, at the first opportunity the gun should be checked and repaired.

b. Gun in battery. (1) *Breechblock not closed.*

(*a*) Close it manually, relay, and fire.

(*b*) If it will not close, see if ammunition is seated. If ammunition will not seat, remove round, reload, relay and fire. If ammunition still will not seat, clean chamber, reload, relay, and fire.

c. Ammunition is seated. Check for broken or worn extractor and closing spring, replacing if necessary. Clean and lubricate bearing surfaces of breechblock if necessary.

(1) *Breechblock is closed.*

(*a*) Recock piece. If cocking action shows that piece is still cocked (indicated by no resistance other than the cocking lever plunger spring), examine for bent or broken trigger arm, and for malfunction of trigger actuator mechanism. Replace parts if possible, relay and fire. If trigger mechanism is not defective, remove firing spring retainer and check for weak or broken firing spring.

(*b*) If cocking action indicates that the sear lug has been released from the sear shoulder (indicated by heavy resistance to cocking action), relay and fire. If gun still fails to fire after twice repeating this action, remove percussion mechanism; clean, lubricate, replace defective parts, recock, relay, and fire.

(*c*) If gun fails to fire, unload, load with a new round, relay and fire.

(*d*) If cocking action shows that the sear lug will not remain engaged on the sear shoulder, disassemble breechblock, clean and replace defective parts.

Section VI

MOUNTED ACTION

19. PREPARE TO FIRE. The crew being at mounted posts, with hatches open.

Tank Commander	Gunner	Bow Gunner	Driver
Command PREPARE TO FIRE.			
Inspect bore and chamber of 37-mm.	Check solenoids and manual firing controls.	Close hatch; raise and clean periscope.	Close hatch; raise and clean periscope.
Half-load coaxial machine gun.	Help load coaxial machine gun.		
Unlock turret traverse lock.	Unlock gun traveling lock and turret friction clamp.	Report "Bog's hatch closed".	Report "Driver's hatch closed".
Clean exterior of Gunner's and Tank Commander's sights.	Wipe off sights. Make field check of sights (see Par 41).	Unlock bow gun traveling lock.	Start auxiliary generator.

Tank Commander	Gunner	Bow Gunner	Driver
	Check manual traverse; engage power traverse. Turn stabilizer on if situation warrants. Close hatch.	Half-load bow gun.	
Command REPORT.	Report "Gunner Ready".	Report "Bog Ready".	Report "Driver Ready".

20. DUTIES IN FIRING. Tank prepared to fire with guns loaded.

Tank Commander	Gunner	Bow Gunner	Driver
Give fire order as prescribed in FM 17-12. Reload 37-mm gun (except when observing from open turret hatch). Inspect all rounds	Fire on targets as ordered by Tank Commander. Reload 37-mm gun if	Fire on designated targets and on emergency targets that appear. When not firing, observe	After fire order is issued, avoid changing course and maintain constant speed if moving. (Steer only to

before loading, and wipe off if necessary.	Tank Commander is observing from open hatch, or if he is in position where he cannot load.	in assigned sector.	avoid obstacles, or if change of direction is ordered; avoid excess steering or small changes). Warn Gunner if about to pass over rough ground or if about to change course by calling ROUGH, or CHANGING COURSE.
Signal ready by tapping Gunner on back. When gunner calls MISFIRE, check that breech is closed and gun is in battery. Help Gunner recock gun.	If necessary, change recoil knob setting. Call MISFIRE if 37-mm fails to fire. Recock gun and fire. If gun fails to fire after twice repeating this action, see Par. 17.		Select speed which can be maintained over the particular terrain. When tank fires from stationary position, continue to run engines unless ordered otherwise.

Tank Commander	Gunner	Bow Gunner	Driver
When Gunner calls STOPPAGE, reduce stoppage in coaxial machine gun. Fire coaxial machine gun at Gunner's command if solenoid fails to operate. Fire antiaircraft machine gun. Control Driver over interphone.	Call STOPPAGE if coaxial machine gun fails to fire. Notify Tank Commander when to fire coaxial machine gun in case solenoid fails to operate. During lulls in firing, observe in assigned sector.		

21. TO SECURE GUNS.

Tank Commander	Gunner	Bow Gunner	Driver
Command CEASE FIRING, SECURE GUNS.			
Clear coaxial MG, unload 37-mm gun, inspect bore and close breech.	Traverse turret to front. Turn off firing switch. Turn off power traverse, engage	Clear Bow MG. Lock bow gun traveling lock. Lower periscope, open hatch.	Turn off auxiliary generator. Lower periscope, open hatch. Raise seat to convoy position.
Lower periscopes, open hatch.			

Refill 37-mm ready racks.	manual traverse. Turn off stabilizer; engage manual elevating gear. Open hatch.	Dispose of empty cartridge cases.	
Dispose of empty cartridge cases.	Help refill 37-mm ready racks. Help dispose of empty cartridge cases.		
Lock turret traversing lock.	Lock friction clamp. Lock gun traveling lock.		
If situation permits, order Bog to swab bore.	Make field check of sights.	If so ordered, dismount, swab bore, and remount. Put on bow machine gun tape muzzle cover.	Put on tape muzzle covers.
Command REPORT.	Report "Gunner Ready".	Report "Bog Ready".	Report "Driver Ready"

35

22. TO LOAD AMMUNITION. Wipe off, load and stow 37-mm gun ammunition with great care to avoid burring the rotating band, or denting the case. (See TM 9-1900). Different lots should be segregated. During periods between firing, ammunition is restowed to the most accessible racks. Test each round of ammunition by loading in gun.

23. TO LOAD ALL WEAPONS. a. The tank gun is loaded on order; this is normally the fire order, but some types of action will dictate loading prior to the appearance of the target. Machine guns are clear until the command PREPARE TO FIRE, when they are half-loaded. When the fire order is given, however, or the unit is deployed for combat, all machine guns are fully loaded. This does not necessarily apply to the AA gun, which is elevated and half-loaded as the tactical situation demands.

b. When the tank commander is not in a position to load (observing from the top of the turret), he commands LOAD SHOT (HE). At the command the gunner loads as directed.

Section VII

DISMOUNTED ACTION

24. TO FIGHT ON FOOT THROUGH HATCHES. a. Crew being at mounted posts, hatches open.

NOTE: In all drills which include manning of submachine gun, it is assumed that the tank is equipped with four submachine guns M3. Therefore, when the cal .30 MG is not dismounted to fight on foot, all crew members will dismount with submachine gun M3, plus a submachine ammunition case containing six 30-round clips.

Tank Commander	Gunner	Bow Gunner	Driver
Command PREPARE TO FIGHT ON FOOT.	Procure grenades as ordered.	Procure grenades as ordered.	
Order distribution of grenades.	Disconnect breakaway plugs.	Disconnect breakaway plugs.	
Disconnect breakaway plugs.		Clear bow gun (if loaded).	
Secure submachine gun and ammunition carrying case.			

Tank Commander	Gunner	Bow Gunner	Driver
Command DISMOUNT.			
Dismount with submachine gun ammunition and binoculars.	Dismount to receive spare parts roll and spare bolt assembly from driver.		Pass spare parts roll and spare bolt assembly to Gunner.
	Receive three boxes cal .30 ammunition; place in front of tank.	Pass three boxes cal .30 ammunition to Gunner.	
Dismount tripod*, lay in front of tank. Pick up two boxes cal .30 ammunition. Lead crew to MG position.	Receive bow MG from Bog.	Dismount bow gun and pass it to Gunner. Dismount from tank, pick up tripod and one box cal .30 ammunition.	Remain in tank; drive it to concealment. Disconnect breakaway plugs. Move to turret; connect breakaway plugs to Tank Commander's control box. Maintain contact with Platoon Leader.

Tank Commander	Gunner	Bow Gunner	Driver
Supervise firing of MG; cover gun crew with submachine gun.	Mount gun.	Mount tripod.	Man AA or 37-mm gun from turret, as situation demands.
	Man gun as #2.	Man gun as #1.	

*Make sure pintle and elevating mechanism are in tripod case. If not, Bog obtains them from spare parts box and passes to Gunner.

b. The dismounted crew moves to the position indicated by the tank commander, or in drill as indicated in figures 12 and 13. The crew members take posts and perform duties of the crew of a ground mounted machine gun as prescribed for gun drill in FM 23-55.

c. In combat it is assumed that the tank will be moved to a concealed position if possible, before the crew dismounts. The driver remains in the tank, moves it to concealment, and mans the antiaircraft machine gun or tank cannon as the situation demands.

25. TO REMOUNT FROM ACTION THROUGH HATCHES.

Tank Commander	Gunner	Bow Gunner	Driver
Command OUT OF ACTION, MOUNT.			
Supervise taking MG out of action.	Dismount gun.	Fold tripod, lay tripod and cal .30 ammunition box in front of tank.	

39

Tank Commander	Gunner	Bow Gunner	Driver
	Pass bow MG to Bog.	Take mounted post. Receive bow MG and mount it.	Resume mounted post.
Strap tripod on tank fender.	Pass spare parts roll and bolt assembly to Driver.		
Mount to turret, taking submachine gun and ammunition.	Pass 3 boxes cal. .30 ammunition to Bog.	Receive and stow 3 boxes cal .30 ammunition from Gunner.	Receive and stow spare parts roll and bolt assembly.
Connect breakaway plugs.	Connect breakaway plugs.	Connect breakaway plugs.	Connect breakaway plugs.
Command REPORT.	Report "Gunner Ready".	Report "Bog Ready".	Report "Driver Ready".

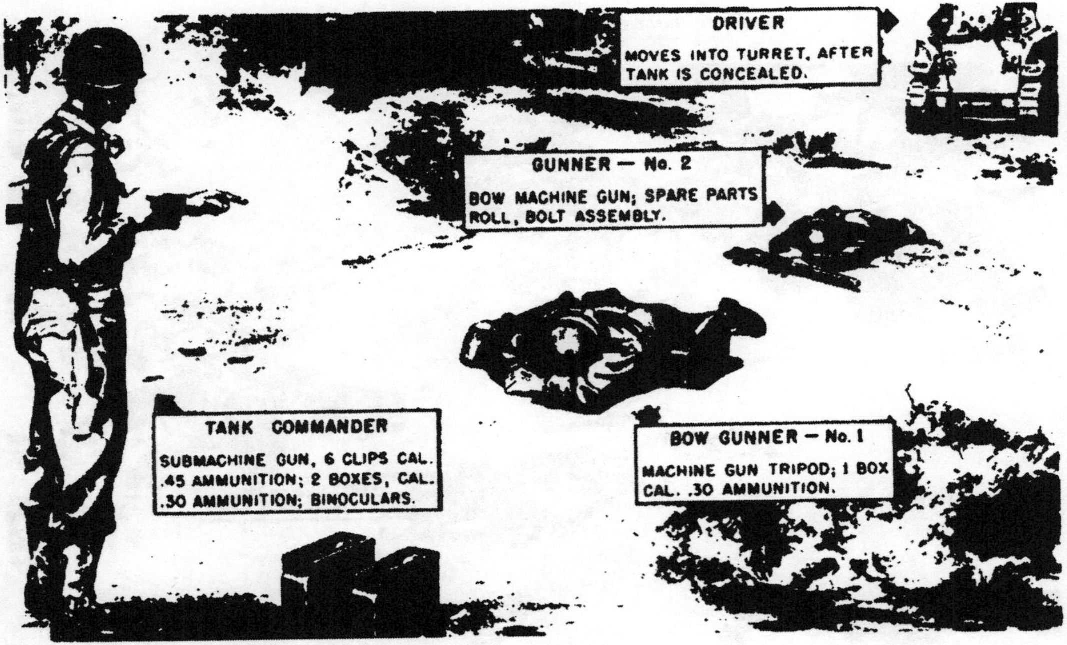

Figure 12. Dismounted action, crew formed for drill (tank commander instructing).

Figure 13. Posts of dismounted crew in action.

26. ACTION IN CASE OF FIRE. a. Fire in engine compartment. The first crew member to discover the fire calls ENGINE FIRE—

Tank Commander	Gunner	Bow Gunner	Driver
Disconnect breakaway plugs		Disconnect breakaway plugs	
Receive "speed" wrench from Bog.	Pull fixed fire extinguisher control handle.	Pass "speed" wrench (from tool box) to Tank Commander.	Stop engines.
Dismount to rear of tank; open engine doors with wrench.	Receive hand extinguisher from Bog. Relay hand extinguisher to Tank Commander.	Pass hand extinguisher to Gunner in turret.	
Receive extinguisher from Gunner, stand by to use it in case fixed extinguisher does not put out fire.	Remain at post for further orders.	Dismount; go to rear of tank and assist Tank Commander.	Remain at post for further orders.

b. Fire in turret or driver compartment. The first crew member to discover the fire calls TURRET (HULL) FIRE. The tank is stopped and engines shut off. Fire extinguishers are passed to men nearest fire and other crew members help them in any way possible to extinguish the flame. The turret is traversed in any direction which will aid crew to reach fire with extinguishers.

43

27. TO ABANDON TANK. If it becomes necessary to abandon tank, the crew proceeds as in paragraph 10, 24, or 26, with the following changes or additions:

a. Time permitting deliberate action, the tank commander displays the flag signal DISREGARD MY MOVEMENTS, and supervises the disabling of those weapons which remain in the tank. Backplates are removed from machine guns and the percussion mechanism from the tank gun. All similar spare parts are also removed. Individual weapons and maximum possible ammunition loads are carried. The driver dismounts in order with the rest of the crew.

b. Ordinarily the tank is abandoned as a result of a direct hit which either causes it to catch fire or disables it so that it becomes a vulnerable target. There may be a time interval of as little as five seconds in which the crew can escape without further injury. At the command ABANDON TANK, crew members throw open hatches, climb out, jump to the ground and take cover at a safe distance from the tank. It is particularly important in case of fire to hold the breath until clear of the vehicle. Inhaling the fumes and smoke of the fire may injure the lungs and will at least incapacitate the individual for a time.

28. ADVICE TO INSTRUCTORS. a. Disciplined and effective dismounted action requires long and arduous drill. Satisfactory results are obtained only by painstaking repetition of each movement.

b. Training in dismounted action is undertaken in the field rather than in the tank park. Crews are required to dismount to fight on foot on all types of terrain and under every variety of simulated combat conditions with full loads of ammunition. Rough terrain complicates the problem of dismounting from the escape hatch and develops ingenuity and physical agility not possible in tank park training.

c. Instructors explain and demonstrate to tank crews how necessary to their safety and success in combat is a high state of training in dismounted action. They must point out that dismounted action from disabled tank taken under small arms fire usually is practicable only from the escape hatch, and that skill and practice in use of the escape hatch will pay dividends. The escape door is kept clean and well lubricated so that its release is immediate and positive. Frequent inspection of the mechanism is made by the tank commander to see that the locking rods are not bent.

29. PARK AND BIVOUAC PRECAUTIONS. a. Always have a guide when moving a tank in park or bivouac.

b. Keep at least 10 feet in front of tank and to one side of its path when directing the tank forward or backward in park or bivouac.

c. Walk, do not run while guiding a tank.

d. Mount and dismount without using tube of gun as a hand hold, or breech end of gun as a foot hold in entering turret.

Section VIII

EVACUATION OF CASUALTIES FROM TANKS

30. GENERAL. a. The drill prescribed in the following paragraphs is for use in training crew members so that any two of them may evacuate casualties with maximum efficiency. A tank disabled by an enemy projectile may expect another hit without delay. When ammunition is ignited, evacuation must be effected in a few seconds. Therefore speed, rather than care in handling, is the primary consideration in removing a casualty from a tank.

b. Evacuation of casualties is undertaken only:

(1) When the tank is disabled.

(2) When the position of the casualty in the tank prevents the crew from functioning.

(3) At the rallying point.

After evacuation, the casualty may be carried to a protected area where emergency first-aid is administered.

31. METHODS AND DEVICES. Not more than two men can work effectively at a single hatch opening. If the man nearest the casualty is unhurt, and tactical considerations permit, he can help by remaining inside and improvising a sling (made up of pistol belts, waist belts, or field bag straps, see FM 17-80), or by moving the casualty to a position where he can be grasped from above and then aid in boosting him out. However, speed will usually dictate that the casualty be grasped by his clothing or by his arms for removal (figures 14 and 15). If there are injuries which will be aggravated by such methods, and time permits, some form of sling may be improvised to relieve that part from further injury (figure 16).

Figure 14. Evacuation of bog with arms crossed.

Figure 15. Evacuation of bog down front slope plate.

142

Figure 16. Evacuation of gunner with two pistol belts.

32. TO EVACUATE GUNNER (TANK COMMANDER). Crew being at mounted posts. (Note: The duties of No. 1 and No. 2 of evacuation team are interchangeable. The first man to realize the situation assumes the initiative as No. 1).

No. 1	No. 2
Command EVACUATE GUNNER (TANK COMMANDER)	
Mount to top of turret.	Throw first-aid kit (kept in accessible spot in turret) to ground.
	Mount to top of turret.
Slip pistol belt under each arm of gunner.	Pass pistol belt to No. 1.
Lift casualty through hatch.	Lift casualty through hatch.

48

No. 1	No. 2
	Hold body in a sitting position on hatch cover.
Lower casualty to left sponson (or protected side, if tank is disabled).	Lower casualty to left sponson (or protected side if tank is disabled).
Jump to ground; support lower portion of body as casualty is moved off tank.	Jump to ground; support upper portion of body as casualty is moved off tank.
Carry casualty to a protected area.	Help No. 1 carry casualty to a protected area.
Report location and condition of casualty.	

33. TO EVACUATE BOG (DRIVER). Crew being at mounted posts.

No. 1	No. 2
Command: EVACUATE BOG (DRIVER).	
Descend through turret opening into Bog compartment, if necessary to open locked hatch cover.	Move to Bog hatch opening.
Throw out first-aid kit.	Straddle 37-mm gun, facing rearward.
Cross arms of casualty; pull one arm through hatch as No. 2 pulls on other arm.	Pull one arm of casualty through hatch as No. 1 pulls on other arm.
Turn body to face rearward as it is lifted out.	Turn body to face rearward as it is lifted out.
	Rest body in a supported position on rim of hatch opening.

No. 1	No. 2
Jump to ground in front of tank; support trunk of body as it falls backward over forward slope of tank.	Allow body to fall back into No. 1's arms. Make sure legs are freed from hatch opening.
Carry casualty to a protected area.	Jump to ground; support lower portion of body as it is carried to a protected area.
Report location and condition of casualty.	

NOTE: In case a large man is to be evacuated, one man of team may have to enter vehicle and assist from the inside.

Section IX

INSPECTIONS AND MAINTENANCE

34. GENERAL. a. The tank commander is responsible for seeing that all inspections are made. He receives reports from the various crew members relative to their individual inspections, and he indicates in his report anything requiring the service of maintenance personnel. In supervising first echelon maintenance he uses his discretion in delegating additional responsibilities to other crew members.

b. Inspection covers all personal equipment and weapons, vehicle equipment and weapons and mechanical features of the vehicle. In combat it includes a check of the application of protective cream by the crew members. Checks of instruments, lights, siren, track, suspension system, and engine performance are made in accordance with provisions of the appropriate technical manual; the driver fills in his Driver's Report indicating required maintenance work. The Driver's Report should be carefully and thoroughly prepared. Any irregularity noted and entered on the report, which is not repaired before the tank is used again, should be re-entered on the report continually until it has been properly taken care of.

35. BEFORE OPERATION INSPECTION. Tank locked and covered with tarpaulin. (NOTE: For training purposes, the inspection is divided into four phases, each phase being completed by all crew members before the next phase is begun. Crew members use tools as needed and report and correct deficiencies as found).

PHASE A

Tank Commander	Gunner	Bow Gunner	Driver
Command FALL IN: PREPARE FOR INSPECTION.			
Inspect crew.	Stand inspection.	Stand inspection.	Stand inspection.
Command PERFORM BEFORE OPERATION INSPECTION.			
Supervise inspection.	Help remove tarpaulin.	Help remove and fold tarpaulin.	Remove and fold tarpaulin (3' x 6')
Fill out trip ticket during inspection.			
Inspect outside equipment.	Mount and unlock tank.	Check for water, fuel and oil leaks around and under tank.	Lay tarpaulin to right of tank.

52

Tank Commander	Gunner	Bog	Driver
Receive and assemble rammer staff from Bog and MG cleaning rod and rags from gunner.	Open turret hatches; enter turret. Check that turret guns are clear. Move to Driver's compartment; unlock Driver's and Bog's hatches. Pass MG cleaning rod and rags to Tank Commander. Pass tools out to Driver. Receive and stow gun covers; clear bow guns. Stow breech covers. Open fuel shut-off valve.	Check radiators and fuel tanks for level. Pass rammer staff to Tank Commander. Check final drives for oil level.	Remove cannon muzzle cover, and bow gun cover. Receive tools from Gunner. Pass gun covers to Gunner. Lay tools on tarpaulin and check them. (Figure 17).
Command REPORT.	Report "Gunner Ready".	Report "Bog Ready".	Report "Driver Ready".

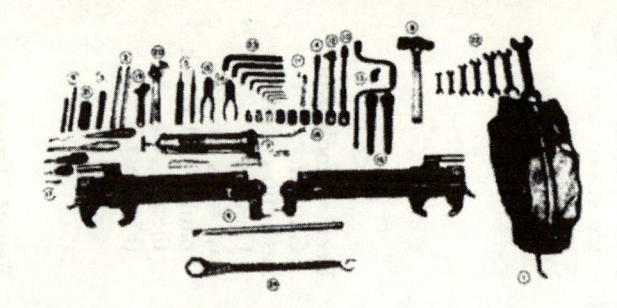

Figure 17. Vehicular tools furnished each light tank (see SNL G-103). (The number and position of tools for a formal inspection is a command decision.)

1. Bag, tool
2. Bars, utility (2)
3. Chisel, machs, hand, cold
4. Extension, socket wrench, plain, 10"
5. Files, A.S., 6" & 8" (2)
6. Fixture, track connecting and link pulling w/handle (commonly called "track jack")
7. Gun, lubr., pressure, hand
8. Hammer, machs, ball peen, 32 oz
9. Handle, cross extension, socket wrench, 8"
10. Handle, hinged socket wrench, 12", w/crossbar
11. Handle, socket wrench, offset, dble-end
12. Handle, "T", sliding, socket wrench
13. Joint, universal, socket wrench
14. Pliers, comb., slip-jt, 8"
15. Pliers, side-cutting, fl-nose, 8"
16. Ratchets, reversible, socket wrench, 9" (2)
17. Screwdrivers, including one non-magnetic for Pioneer compass (5)
18. Sockets, 1/2" sq-drive, dble-hex, 7/16" to 3/8" (9)
19. Wrench, adj., sgle-end, 8"
20. Wrench, adj., sgle-end, 12"
21. Wrench, brake, socket, w/handle
22. Wrenches, engrs, dble-hd, 5/16" to 1-3/8" (7)
23. Wrenches, socket head set screws, 1/8" hex to 3/4" hex (10)
24. Wrench, trailing idler wheel nut & track adjusting
25. Wrench, socket, speeder

54

PHASE B

Tank Commander	Gunner	Bow Gunner	Driver
Command PERFORM PHASE B. Assist Gunner in sight adjustment.	Make sight adjustment. See paragraph 42.	Check with Driver that master switch is off.	Enter Driver's seat. Check that master switch is off.
Swab bore of all guns. Put tape muzzle covers on all guns.	Open gun breech. Check following: Coaxial and AA MG and adjust headspace. 37-mm ammunition. Cal .30 ammunition. 37-mm and MG mounts (close breech).	Open engine compartment doors. Check following: Trigger type oil can. Engine oil levels. Engine compartment for water, oil and fuel leaks. Accessories and drives for adjustment and security.	Inspect and check following: Battery. Oil cans and oil for vehicle. Transfer unit and differential oil level. Steering levers, parking brakes. Transmission and transfer levers for operation.

Tank Commander	Gunner	Bow Gunner	Driver
		Air cleaners and connections. Close engine compartment doors.	Auxiliary generator oil level. Start generator; check operation.
Command REPORT.	Report "Gunner Ready".	Report "Bog Ready".	Report "Driver Ready".

PHASE C

Tank Commander	Gunner	Bow Gunner	Driver
Command PERFORM PHASE C.		Notify Driver to start engines.	When notified by Bog, turn on master switch.
Inspect right track and suspension system. Direct Driver to move forward one tank length.	Check following: Oil in turret reservoir. Oil can and stabilizer oil.	Inspect left track and suspension system.	Start engines. Check instruments, warning lights and siren. Move tank as directed by Tank Commander.
Check right support rollers, bogie	All firing controls.	Check left support rollers, bogie	

56

wheels, idler and sprocket as tank moves forward.	Turn on turret master switch.	wheels, idler wheel and sprocket as tank moves forward.
Inspect that part of track not visible before.		Inspect that part of track not visible before.
Check for tightness of wedge nuts and sprocket ring cap screws.	Check stabilizer operation, oil reservoir connections and pump.	Check for tightness of wedge nuts and sprocket ring cap screws.
Replace tools in bag and pass bag to Driver.		
		Assist Driver in checking driving and blackout lights.
		Check MG tripod case for pintle and elevating mechanism.
		Install radio antenna.
		Replace tarpaulin on rear deck.
		Stop engines.
		Receive and stow tools.
		Check driving and blackout lights.

57

Tank Commander	Gunner	Bow Gunner	Driver
Pass rammer staff to Bog and cleaning rod to Gunner. Mount to turret.	Receive and stow machine gun cleaning rod. Check with Driver and Bog that their hatches are closed. Check manual elevation, operation of hand and power traverse, and turret lock.	Receive and stow rammer staff. Mount and close hatch.	Close hatch.
Assist in checking recoil oil. Connect breakaway plugs. Command REPORT (interphone check)	Check recoil cylinder oil. Connect breakaway plugs. Report "Gunner Ready".	Connect breakaway plugs. Report "Bog Ready".	Connect breakaway plugs. Report "Driver Ready".

NOTE: In later models of the Light Tank M5A1, there is an *emergency ignition switch* on the right side of the hull roof. This switch is left in the "ON" position except in emergencies, when it is used by the Tank Commander or other personnel within reach.

58

PHASE D

Tank Commander	Gunner	Bow Gunner	Driver
Command PERFORM PHASE D. Check following: Gun books, Manuals, Accident Forms, Check Chart, Driver's permit, Periscope, spare, and spare head. Safety belt. Make first echelon check of radio and interphone system. (See Par. 41)	Check transmission oil levels. Notify Driver to start engines. Check following: Safety belt, Hull compass, Periscope, spare, spare heads, Grenades, Presence of spare parts, tools and accessories for all guns.	Check bow gun and adjust headspace; put on cartridge bag. Check following: Bow MG ammunition, Periscope, spare, and spare head. Flag set, Fire extinguishers, Decontaminating apparatus. Vehicular first-aid kit, Crew rations. Escape hatch operation, Water cans, Canvas bucket,	Open hatch. Check following: Cooking stove, Periscope, spare, and spare head. Safety belt. Start engines when notified by Gunner. Constantly observe operation of engines for smoothness, synchronization, unusual noises.

Tank Commander	Gunner	Bow Gunner	Driver
Command REPORT.		Safety belt. Open hatch	
	Report "Gunner Ready".		
		Report "Bog Ready".	
			Report "Driver Ready".
Complete trip ticket. Report READY to Platoon Leader.			

NOTE: The flame thrower, on tanks so equipped, is checked in this phase. The crew member using the weapon checks its condition, mechanism, and the fuel level in its tank in accordance with the appropriate published guide. Where it is used as an alternate weapon to the bow machine gun, it is mounted on order of the tank commander.

36. INSPECTION DURING OPERATION. This is a continuous process of checking by all crew members.

Tank Commander	Gunner	Bow Gunner	Driver
Remain alert for unusual noises or conditions. Assist Driver to avoid	Check following: Security of turret lock. Security of gun.	Watch instruments and warning lights. Listen for unusual noises.	Check all instruments carefully. Check controls. Listen for unusual noises.

obstacles that would cause injury to tank or crew.
Check radio interphone system, and security of radio antenna.
Check security of AA gun, stowage, and equipment.

| Stabilizer operation. | Check security of bow gun. | |

37. INSPECTION AT THE HALT. The length of halt determines the extent of this inspection. Items listed below are in order of normal priority.

Tank Commander	Gunner	Bow Gunner	Driver
Command PERFORM HALT INSPECTION. Supervise halt inspection (if required to dismount the Gunner must stand by AA gun.)	Clean sights and periscopes. Make field check of sights (par 41). Check following: Security of guns and mounts.	Disconnect breakaway plugs. Dismount and close hatch. Check final drives for leaks, excessive temperatures.	Disconnect breakaway plugs. Idle engine 4-5 minutes before stopping. Check engine synchronization and unusual noises.

61

Tank Commander	Gunner	Bow Gunner	Driver
	Stabilizer connections. Turret mechanism. (Check that bow hatches are closed.)	Check under tank for water, oil, or fuel leaks. Check radiators for level, leaks. Check fuel tanks for level and leaks.	Check instruments, warning lights, parking brakes. Cut engines, close hatch and dismount. Open engine compartment doors. Make visual check of engine compartment for fuel, water and oil leaks, security of accessories. Check engine oil levels.
Stand by AA gun.	Radio antenna. Security of equipment in general. Place gun in traveling position. Lock turret lock. Check stowage of equipment in turret.	Check presence and security of outside equipment. Check and clean suspension system. Check presence and security of outside equipment. Help check air cleaners. Take mounted post. Clean periscope. Connect breakaway plugs.	Close engine compartment doors. Check air cleaners. Mount Driver's seat. Clean periscope. Connect breakaway plugs.

Tank Commander	Gunner	Bow Gunner	Driver
Command REPORT.	Report "Gunner Ready".	Report "Bog Ready".	Report "Driver Ready".

Report READY to Platoon Leader.

38. AFTER OPERATION MAINTENANCE. a. After the operation the tank is immediately given whatever servicing and maintenance is needed to prepare it in every way for further sustained action. *This servicing covers all the points listed in the Before Operation Inspection and covers them in the same order, with obvious modifications.* (For example, the tank is locked at the end of the inspection instead of being unlocked at the beginning; the check for leaks under the tank is more effective after it has stood for a while; battery switches are turned off rather than on and only after all checks requiring use of battery power; equipment is covered and stowed rather than being uncovered and made ready for use).

b. The tank will be completely cleaned, serviced, and replenished (fuel, oil—all types, grease, coolant, ammunition—all types, first aid kit, water, and rations). *All special precautions against fire will be observed while refueling.* Crew members will perform the following additional operations not covered in the Before Operation Inspection.

Tank Commander	Gunner	Bow Gunner	Driver
Command PERFORM AFTER OPERATION MAINTENANCE.	Clean weapons.		
Forward completed trip ticket to platoon leader, and report of all necessary 2d echelon maintenance, fuel, lubricants, ammunition and rations required.		Help Driver clean tank.	Idle engines 4 to 5 minutes before stopping.
			Clean tank suspension and outside of tank.
		Help Gunner clean weapons.	Help Gunner clean weapons.

39. PERIODIC ADDITIONAL SERVICES. NOTE: In garrison these services are performed weekly; on maneuvers or in combat they are performed after each field operation.

Tank Commander	Gunner	Bow Gunner	Driver
Command FALL IN: PREPARE FOR INSPECTION.			
Inspect crew.	Stand inspection.	Stand inspection.	Stand inspection.

Command			
PERFORM PERIODIC INSPECTION.	Enter turret.	Tighten all wedge nuts, and inspect track and suspension system.	Enter Driver's compartment. Clean battery and case thoroughly. Make hydrometer reading. Check water level. Charge if necessary.
	Tighten all loose bolts, nuts and connections in turret.	Tighten all loose bolts, nuts and connections outside tank.	Tighten all loose bolts, nuts and connections in compartment.
Inspect and supervise work of crew members.	Lubricate gun and mount as needed. Check and clean parts and tools for weapons. Drain sediment from fuel tanks.	Clean fuel filter. Tighten all loose bolts, nuts, and connections in compartment. Clean compartment. Clean and oil escape hatch.	Clean compartment. Operate and check floor drain valve.

65

Tank Commander	Gunner	Bow Gunner	Driver
	Help perform 250-mile lubrication. Clean and touch-up any rust spots in turret.	Operate and check floor drain valve. Help perform 250-mile lubrication.	Dismount; open engine doors; clean engine and engine compartment; operate and check engine compartment drain valves. Perform 250-mile lubrication, referring to appropriate guide. Close engine doors.
Command REPORT.	Report "Gunner Ready".	Report "Bog Ready".	Report "Driver Ready".

40. RADIO. The tank commander will make the following inspection of radio sets (SCR-508, SCR-528, and SCR-538) prior to operation.

a. Cords. See that insulation and plugs are dry, unbroken, and making good contact. Arrange loose cordage to prevent its entangling personnel or equipment.

b. Antenna. See that

(1) Mast sections are tight. (Do not remove taped joints.)

(2) Leads at transmitter, receiver, and mast base are tight.

(3) Mast base is tight and not cracked.

(4) Insulators that pass through the armor plate and bulkheads are not broken or out of place.

c. Radio set mounting, snaps, snubbers, etc. Check for security and condition.

d. Microphones, switches, headsets. Check for condition and proper positions. Replace all defective headsets and microphones from spares and turn defective equipment in for repair or replacement.

e. Spare antenna sections. See that they are correctly placed in the roll and stowed to avoid being damaged.

f. Set grounding. Check connections.

g. Tubes. See that they are firmly seated in the sockets. Turn in the defective tubes at the earliest opportunity.

h. Fuses. Check condition, and spare supply for number and proper rating.

i. Cleanliness. See that radio sets and associated equipment are clean.

j. Battery voltage. Check with driver to see that the battery voltage is kept up. If voltage is low, have driver start the auxiliary generator. Always have generator running when radio is used and tank engine is not.

k. Crystals. Check for number, position, and frequency. Be sure required crystals are present.

Section X

SIGHT ADJUSTMENT

41. FIELD CHECK OF SIGHTS. Frequent checking of the sights is vital in the field. This is done by selecting a distant aiming point as explained in paragraph 42a, and aligning the crosshairs of the coaxial telescope on it. Then check to see whether the crosshairs of the periscopic telescope coincide on the same point. If they do, the sights are assumed to be still in adjustment. If they do not, the adjustment has slipped and a new adjustment for both sights must be made.

42. SIGHT ADJUSTMENT. Sights are in adjustment when the axis of the bore is parallel to the line of sighting of the telescope (and periscope). For practical purposes these lines are considered parallel when they converge on a point not less than 1500 yards distant.

a. Distant aiming point method. Select a point such as a building, telegraph pole, or smokestack at least 1500

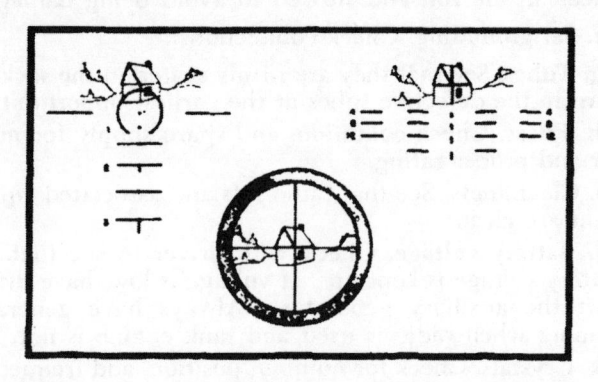

Figure 18. Sight adjustment on distant aiming point.

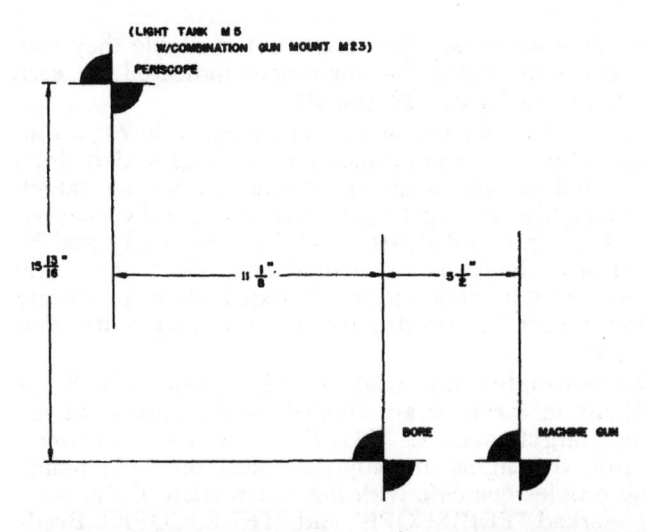

Figure 19. Testing target for combination gun mount M23.

yards away, preferably one having distinct vertical and horizontal straight lines which intersect, and thus render the alignment of the crosshairs easier and more accurate. Line the gun tube on this point, using the issue bore sights if they are available, or improvising crosshairs for the muzzle and sighting through the firing pin well of the closed breechblock. Then move both sights vertically and horizontally until their crosshairs coincide with the selected aiming point (figure 18). When there is a clamping device to secure this adjustment of the coaxial telescope once it is made, recheck after tightening to be sure that the adjustment has not been thrown off. Readings of the adjusting knob indices are recorded on all spare heads of the periscopic sights.

b. Testing target. When vision is limited by atmospheric conditions, close country, or jungle growth, the best field check for sighting equipment is the testing target, since it requires a clearing only 80 to 120 feet long. If issue testing targets are not available they may be constructed using the dimensions indicated for each gun mount in figures 19 and 20.

NOTE: Manufacturing variance frequently causes individual tanks to have dimensions varying widely from those used on the issue or constructed testing target. Therefore, the testing target must be carefully checked against the gun and sights which have been adjusted by the distant aiming point method on a clear day. Any discrepancies in dimensions will require that a separate testing target be constructed for use *only with that tank gun.*

The boresights are installed as in a, after which the crosshairs in the tube are aligned on the portion of the testing target marked "BORE". Without disturbing the gun, the sights are adjusted until the zero points in the reticles coincide with the intersection of the cross lines marked "PERISCOPE" and "TELESCOPE". Readings of the adjusting knob indices are recorded on all spare heads of the periscopic sights.

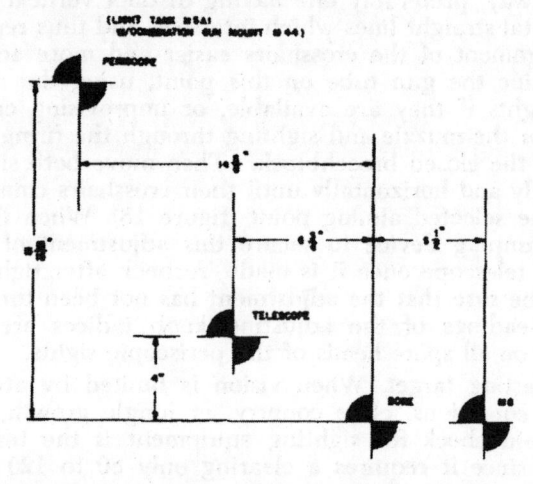

Figure 20. Testing target for combination gun mount M44.

Section XI

DESTRUCTION OF MATERIEL AND EQUIPMENT

43. GENERAL. a. The destruction of materiel is a command decision to be implemented only on authority delegated by the division or higher commander. This is usually made a matter of standing operating procedure. *It is ordered only after every possible measure for preservation or salvage of the materiel has been taken, and when in the judgment of the military commander concerned such action is necessary to prevent—*

(1) Its capture intact by the enemy.

(2) Its use by the enemy, if captured, against our own or allied troops.

(3) Its abandonment in the combat zone.

(4) Knowledge of its existence, functioning, or exact specifications from reaching enemy intelligence.

b. The principles followed are

(1) Methods for the destruction of materiel subject to capture or abandonment in the combat zone must be adequate, uniform, and easily followed in the field.

(2) Destruction is as complete as available time, equipment, and personnel permit. If thorough destruction of all parts cannot be completed, the most important features of the materiel are destroyed, and parts essential to the operation or use of the materiel and which cannot be easily duplicated, are ruined or destroyed. *The same essential parts are destroyed on all like units to prevent the enemy's constructing one complete unit from several damaged ones by "cannibalism".*

c. Crews are trained in the prescribed methods of destruction. *Training does not involve the actual destruction of materiel.*

d. Methods. (1) The methods below are given in order of effectiveness. If method No. 1 cannot be used, destruction is accomplished by one of the other methods outlined in order of priority shown. Adhere to the sequences.

(2) Certain methods require special tools and equipment such as TNT and incendiary grenades, which may not be items of issue normally. The issue of such special tools and materiel, the vehicles for which issued, and the conditions under which destruction will be effected are command decisions in each case, according to the tactical situation.

44. DESTRUCTION OF THE 37-MM GUN, TANK. a Remove the periscopic and telescopic sights. *If evacuation is possible, carry the sights.* If evacuation is not possible, thoroughly smash the sight and all spare sights.

b. Method No. 1. (1) Open drain plug on recoil mechanism, allowing recoil fluid to drain. It is not necessary to wait for the recoil fluid to drain completely before igniting the fuse in (5) below.

(2) Remove an HE shell from a complete round and seat the shell in the chamber.

(3) Plug the bore for approximately two-thirds of its length, using a rammer staff wrapped with cloth or waste to make it fit tightly in the bore. Mud, stones, clay, or other material may be used to plug the bore in lieu of the cleaning staff.

(4) Cut down a 1/2-pound TNT block to fit snugly in the chamber behind the HE shell. Insert a tetryl nonelectric cap into the TNT block with approximately 3 or 4 feet of safety fuse. Close the breech as far as possible without damaging the safety fuse.

(5) Ignite the safety fuse and take cover. Elapsed time: 2 or 3 minutes if rammer staff is used to plug the bore and the cut-down TNT block is carried with gun; longer if other bore obstructions are used.

c. **Method No. 2.** (1) See **b** (1) above.

(2) See **b** (3) above.

(3) Insert one complete HE round into gun and close breech.

(4) Take cover and fire the gun, using a cord. Elapsed time: 1 to 2 minutes using cleaning staff to plug the bore; longer if bore is plugged with mud, or other material.

d. **Method No. 3.** From point-blank range fire AP ammunition at the gun tube until it is rendered useless.

e. **Method No. 4.** (Elapsed time: 2 or 3 minutes.)

(1) See **b** (1) above.

(2) Fire one HE round against a similar round jammed in the muzzle. Take precautions prescribed in **c** (4) above.

45. DESTRUCTION OF THE GYRO-STABILIZER.

a. Drain oil from system.

b. Smash oil lines.

c. Smash control box.

d. Place an M14 incendiary grenade on control box and pull pin.

46. DESTRUCTION OF CALIBER .30 MACHINE GUN.
a. **Method No. 1.** Field strip. Use barrel as a sledge. Raise cover and smash down toward front. Deform and break backplate; deform T-slot. Wedge lock frame, back down, into top of receiver between top plate and extractor cam; place chamber end of barrel over lock frame depressors and break off depressors. Insert barrel extension into back of receiver allowing the shank to protrude; knock off shank by striking with barrel from the side. Deform and crack receiver by striking with barrel at side plate corners nearest feedway. Elapsed time: 2-1/2 minutes.

b. **Method No. 2.** Insert bullet point of complete round into muzzle and bend case slightly, distending mouth of case to permit pulling of bullet. Spill powder from case, retaining sufficient powder to cover the bottom of case to a depth of approximately 1/8-inch. Reinsert pulled bullet, point first, into the case mouth. Load and fire this round with the reduced charge; the bullet will stick in the bore. Chamber one complete round, lay weapon on ground, and fire with a 30-foot cord. Use the best available cover as this means of destruction may be dangerous to the person destroying the weapon. Elapsed time: 2 to 3 minutes.

c. Small arms cannot be adequately damaged by firing with the barrel stuck in the ground, with or without a bullet jammed in the muzzle.

d. Machine gun tripod mount, caliber .30 M2. Use machine gun barrel as a sledge. Deform traversing dial. Fold rear legs, turn mount over on head, stand on folded rear legs, knock off traversing dial locking screw, pintle lock, and deform head assembly. Deform folded rear legs so as to prevent unfolding. Extend elevating screw and bend screw by striking with barrel; bend pintle yoke. Elapsed time: 2 minutes.

47. DESTRUCTION OF THE LIGHT TANK, M5 SERIES. a. Method No. 1.
(1) Remove and empty the portable fire extinguishers. Smash radio (see paragraph 50). Puncture fuel tanks. Use fire of caliber .50 machine gun, or a cannon, or use a fragmentation grenade for this purpose. Place TNT charges as follows:

(a) 3 pounds between engines.

(b) 2 pounds against left side of transfer unit as near differential as possible.

(c) One-half pound against left fuel tank. Use only a cap (no fuse) in this charge. Point cap end toward 3-pound charge. Insert tetryl nonelectric caps with at least 5 feet of safety fuse in each charge. Ignite the fuses and take cover. Elapsed time: 1 to 2 minutes if charges are prepared beforehand and carried in the vehicle.

(2) If sufficient time and materials are available, additional destruction of track-laying vehicles may be accomplished by placing a 2-pound TNT charge about the center of each track-laying assemblage. Detonate those charges in the same manner as the others.

(3) If charges are prepared beforehand and carried in the vehicle, keep the caps and fuses separated from the charges until used.

b. Method No. 2. Remove and empty the portable fire extinguishers. Smash radio. Puncture fuel tanks. Fire on the vehicle using adjacent tanks, antitank or other artillery, or antitank rockets or grenades. Aim at the engine, suspension, and armament in the order named. If a good fire is started, the vehicle may be considered destroyed. Elapsed time: About 5 minutes per vehicle. Destroy the last remaining vehicle by the best means available. Danger from cannibalism is inherent in this method.

48. DESTRUCTION OF AMMUNITION. a. General. Time will not usually permit the destruction of all ammunition in forward combat zones. When sufficient time and materials are available, ammunition is destroyed as indicated below. At least 30 to 60 minutes are required to destroy adequately the ammunition carried by combat units. In general, the methods and safety precautions outlined in TM 9-1900 are followed whenever possible.

b. Unpacked complete round ammunition. (1) Stack ammunition in small piles. (Small arms ammunition may be heaped.) Stack or pile most of the available gasoline in cans and drums around the ammunition. Throw onto the pile all available inflammable material, such as rags, scrap wood, and brush. Pour the remaining gasoline over the pile. Sufficient inflammable material is used to insure a very hot fire. Ignite the gasoline and take cover.

(2) 37-mm ammunition is destroyed by sympathetic detonation, using TNT. Stack ammunition in two stacks, about three inches apart, bases toward each other. Use

one pound of TNT to six or seven pounds of ammunition. From cover detonate all TNT charges simultaneously.

c. Packed complete round ammunition. (1) Stack the boxed ammunition in small piles. Cover with all available inflammable materials, such as rags, scrap wood, brush, and gasoline in drums or cans. Pour other gasoline over the pile. Ignite and take cover. (Small arms ammunition must be broken out of boxes or cartons before burning.)

(2) The destruction of packed complete round ammunition by sympathetic detonation with TNT is not advocated for use in forward combat zones. To insure satisfactory destruction involves putting TNT in alternate boxes of ammunition, a time-consuming job.

(3) In rear areas or fixed installations, sympathetic detonation may be used to destroy large ammunition supplies if destruction by burning is not feasible. Stack the boxes, placing in alternate boxes in each row sufficient TNT blocks to insure the use of one pound per six or seven rounds of 37-mm ammunition. Place the TNT blocks at the base end of the rounds. Detonate all TNT charges simultaneously. See FM 5-25 for details of demolition planning and procedure.

d. Miscellaneous. Grenades, antitank mines, and antitank rockets are destroyed by the methods outlined in **b** and **c** above for complete rounds. The amount of TNT necessary to detonate these munitions is considered less than that required for detonating artillery shells.

49. FIRE CONTROL EQUIPMENT. All fire control equipment including optical sights and binoculars, is difficult to replace. It is the last equipment to be destroyed, if there is any chance of personnel being able to evacuate. If evacuation of personnel is made, all possible items of fire control equipment are carried. If evacuation is impossible, fire control equipment is thoroughly destroyed by smashing and burying.

50. DESTRUCTION OF RADIO EQUIPMENT. a. Books and papers. Instruction books, circuit and wiring diagrams, records of all kinds for radio equipment, code books, and registered documents are destroyed by burning.

b. Radio sets. (1) Shear off all panel knobs, dials, etc., with an axe. Break open set compartment by smashing in the panel face, then knock off the top, bottom, and sides. The object is to destroy the panel and expose the chassis.

(2) On top of the chassis strike all tubes and circuit elements with the axe head. On the underside of the chassis, if it can be reached, use the axe to shear or tear off wires and small circuit units. Break sockets and cut unit and circuit wires. Smash or cut tubes, coils, crystal holders, microphones, earphones, and batteries. Break mast sections and break mast base at the insulator.

(3) When possible, pile up smashed equipment, pour on gas or oil and set it on fire. If other inflammable material, such as wood, is available, use it to increase fire effect. Bury smashed parts whether burned or not.

APPENDIX

REFERENCES

See FM 21-6, *List of Publications for Training*, and FM 21-7, *List of Training Film Strips, and Film Bulletins*, for full list of references.

FM 17-5	Armored Force Drill.
FM 17-12	Tank Gunnery.
FM 21-5	Military Training.
FM 23-40	Thompson Submachine Gun, Cal .45, M1928A1.
FM 23-41	Submachine Gun, Cal .45, M3.
FM 23-55	Browning Machine Gun, Cal .30.
TM 9-250	37-mm Tank Gun M6.
TM 9-732	Light Tanks M5 and M5A1.
TM 9-850	Cleaning, Preserving and Lubricating Materials.
TM 11-600	Radio Sets SCR-508, SCR-528, and SCR-538.

War Department Lubrication Order No. 81.

Preliminary Inspection for Radio Sets SCR-508, SCR-528, and SCR-538.

SNL G-103	Light Tank, M5 Series.
SNL K-1	Cleaning, Preserving, and Lubricating Materials.
SNL A-45	37-mm Gun M6.
Training Film 17-576	Tank Driving, Part II—Advanced.
Film Strip 2-18	Cavalry Weapons — Browning Machine Gun, Cal .30, M1919A4—Headspace Adjustment, Care and Cleaning, Mechanical Functioning.

78

Film Strip 7-60	Browning Machine Gun, Cal .30, HB, M1919A4 (Ground). Part I.
Film Strip 7-61	Browning Machine Gun, Cal .30, HB, M1919A4 (Ground). Part II.
Film Strip 7-63	Browning Machine Gun, Cal .30, M1917, Part VIII. Section I, Stoppages and Immediate Action.
Film Strip 17-2	Thompson Submachine Gun, Cal .45, M1928A1—Mechanical Training.
Film Strip 17-32	Cal .45 Submachine Gun, M3.

INDEX

	Paragraphs	*Pages*
Abandon tank	27	44
Action:		
Dismounted	24–29	37
In case of fire	26	43
Mounted	19–23	31
Advice to instructors	28	44
Ammunition:		
Destruction	48	75
Inspection before firing	20, 35	32, 52
Loading in tank	22	36
Auxiliary generator:		
To inspect	35	52
To start or stop	19, 21	31, 34
Breechblock:		
To close	15	19
To open	15	19
Crews:		
Composition	3	4
Control	5, 6	6, 8
Drill	7–12	10
Formations	4	5
To close and open hatches	9	13
To dismount	10	14
To form	7	10
To mount	8	11
Destruction of materiel	43–50	71
Dismounted action	24–29	37
Dismounting:		
Through escape hatch	11	15
Through hatches	10	14
Duties in firing	20, 23	32
Empty cases	21	34, 36
Equipment, destruction of	43–50	71
Escape hatch, to dismount through	11	15
Evacuation of casualties	30–33	46
Extinguishers, fire	26	43

	Paragraphs	Pages
Field check of sights	41	68
Fire:		
Engine	26	43
Hull	26	43
Fighting on foot	24, 25	37, 39
Firing:		
Duties of crew	20	32
To cease	21	34
Formations	4	5
Gun crew:		
Composition	13	19
Positions	4, 14	5, 19
Gun, 37-mm tank:		
Destruction	44	72
Firing	15	19
Immediate action	18	29
Inspections	35–39	52
Laying	15	19
Loading and unloading	15	19
Malfunctions and remedies	17	27
Operation	15	19
To secure	21	34
Immediate action	18	29
Inspections:		
After operation	38	63
At the halt	37	61
Before operation	35	52
During operation	36	60
Periodic additional services	39	64
Radio	40	67
Interphone operation	5, 6	6, 8
Maintenance after operation	38	63
Mounted action	19–23	31
Mounting through hatches	8	11
Muzzle covers	21, 35	34, 52
Operation of gun	15	19
Out of action	25	39
Park and bivouac precautions	29	45

	Paragraphs	Pages
Pep drill	12	17
Posts:		
Dismounted	4	5
Mounted	4	5
Prepare to fire	19	31
Purpose and scope	1	1
Radio:		
Destruction	50	77
Inspection	40	67
Operation	5	6
SCR-506	5	6
Ready rack, to refill	21, 22	34, 36
Safety precautions	16, 29	27, 45
Service of the piece	13–18	19
Sight adjustment	41, 42	68
Tank, destruction of	47	74
Turret control	6	8
Weapons:		
Destruction	27, 44, 46	44, 72, 73
Inspection	19, 35	31, 52
To load	23	36

WAR DEPARTMENT FIELD MANUAL
FM 17—76

CREW DRILL AND SERVICE OF THE PIECE MEDIUM TANK, M4 SERIES

(105-MM HOWITZER)

WAR DEPARTMENT — 15 SEPTEMBER 1944

RESTRICTED *DISSEMINATION OF RESTRICTED MATTER.* —The information contained in restricted documents and the essential characteristics of restricted material may be given to any person known to be in the service of the United States and to persons of undoubted loyalty and discretion who are cooperating in Government work, but will not be communicated to the public or to the press except by authorized military public relations agencies. (See also par. 23b, AR 380–5, 15 Mar 1944.)

United States Government Printing Office
Washington 1944

WAR DEPARTMENT
Washington 25, D.C. 15 September 1944

FM 17–76, Crew Drill and Service of the Piece Medium Tank, M4 Series (105-mm Howitzer), is published for the information and guidance of all concerned.

[A.G. 300.7 (15 September 1944)]

By Order of the Secretary of War:

G. C. MARSHALL,
Chief of Staff.

Official:

J. A. ULIO,
Major General,
The Adjutant General.

Distribution:

D 17 (10); R 17(5); I Bn 7 (25), 17 (30)
Interested Battalions
7   T/O & E 7-25
17   T/O & E 17-25

CONTENTS

I. General	1
II. Crew Composition and Formations	2
III. Crew Control	4
IV. Crew Drill	11
V. Service of the Piece	19
VI. Mounted Action	27
VII. Dismounted Action	36
VIII. Evacuation of Wounded from Tanks	49
IX. Inspections and Maintenance	92
X. Destruction of Equipment	97

A Sherman tank of the Sherbrooke Fusiliers covering soldiers of the Fusiliers Mont-Royal as they advance along a street in Falaise, August 1944. (National Archives Canada)

CREW DRILL AND SERVICE OF THE PIECE MEDIUM TANK, M4 SERIES*

(105-MM HOWITZER)

Section I

GENERAL

---

1. PURPOSE AND SCOPE. This manual is designed to present instructional material for the platoon leader and tank commander in training the members of the crew of the medium tank with 105-mm howitzer for combat. It is to be used as a guide to achieve orderly, disciplined, efficient execution of mounted and dismounted action, and precision, accuracy, and speed in the service of the piece. It provides a logical and thorough routine for all inspections of the vehicle and its equipment.

2. REFERENCES. See FM 21-6, FM 21-7, and FM 21-8.

---

* For military terms not defined in this manual see FM 20-205.

Section II

CREW COMPOSITION
AND FORMATIONS

---

3. COMPOSITION. The medium tank crew is composed of five members:

Tank commander _____(LIEUTENANT or
                         SERGEANT)
Gunner _____(GUNNER)
Bow Gunner (assistant
driver (radio operator in
tanks equipped with
SCR-506)) _____(BOG)
Tank driver _____(DRIVER)
Cannoneer (loader and
assistant gunner) (tends
voice radio) _____(LOADER)

4. FORMATIONS. a. Dismounted posts. The crew forms in one rank. The tank commander takes post two yards in front of the right track, facing the front. The gunner, bow gunner, driver, and cannoneer, in order, take posts on the left of the tank commander at close interval.

b. Mounted posts. The crew forms mounted as follows:

(1) *Tank commander.* In the turret, standing on the floor, or sitting or standing on the rear turret seat.

(2) *Gunner.* On the gunner's seat, on the right of the gun.

(3) *Bow gunner.* In the bow gunner's seat.

(4) *Driver.* In the driver's seat.

(5) *Cannoneer.* Standing in the turret, or sitting on the cannoneer's seat at the left of the gun.

Section III

CREW
CONTROL

5. OPERATION OF INTERPHONE AND RADIO.
a. The crew must practice continually with the interphone to obtain its maximum value during combat. It will be used for tank control during operation of the vehicle, radio operation being interrupted during that time.

(1) Helmets and microphones should be worn at all times during crew drill. As standard operating procedure, after mounting, headsets and microphones are tested according to the following procedure:

(a) Cannoneer. 1. Turns OFF-ON switch of radio receiver to ON. (See TM 11-600 for operation of radio and interphone.)

2. Turns OFF-ON switch of transmitter (SCR-508, SCR-528) to ON. (Allow 30 seconds for tubes to warm.)

3. Pushes button of the assigned channel number until it locks.

(b) Crew members. Each crew member inserts the plug of the short cord, extending from his earphones, into the breakaway plug of the headset extension cord of his interphone control box. The microphone is fastened securely in its proper position on the throat or lip to produce maximum clarity of transmission. The microphone is connected to the breakaway plug on the microphone cord of the control box.

(c) Commander. 1. The tank commander depresses the switch on his microphone cord, and orders,

Figure 1. Medium tank, M4 (105-mm howitzer)—front view.

CHECK INTERPHONE. (NOTE: This command is used when the crew mounts by any other method than the drills given in paragraph 8 or 24. In those drills the "Ready" report constitutes the interphone check.) Each member of the crew in the following order: gunner, bow gunner, driver, cannoneer, throws his radio-interphone switch to INT, depresses his microphone switch and reports: BOG CHECK, LOADER CHECK, etc. Upon completion of his report, he immediately returns his switch to RADIO. During this procedure, each crew member adjusts the volume control on his interphone control box to the desired level. *Care must be taken that the microphone switch does not remain in the locked position.* Likewise, the electric cords and the suspension strap must not be wrapped around the hand switch lest they press down on the switch button and cause the dynamotor to burn out.

2. Upon completion of the interphone check at the end of the Before Operation Inspection, or during combat at the last opportunity before the imposition of radio silence, the tank commander tests the operation of the tank radio within the net. To do this he turns his radio-interphone switch to RADIO and either waits for the platoon net to be opened by the NCS or, if the net is open, reports that the Before Operation Inspection is complete.

(2) *Control box positions.* Interphone control box positions are as follows:

(a) *Driver.* On blower bracket above transmission.

Figure 2. Medium tank, M4 (105-mm howitzer)—side view.

(*b*) *Bow gunner.* On blower bracket above transmission.

(*c*) *Gunner.* On right wall of turret to his right.

(*d*) *Tank commander.* On right wall of turret next to gunner's control box. He controls his transmission by manipulating the switch on his control box, marked RADIO-INT, to the type of transmission desired.

(*e*) *Cannoneer.* On left wall of turret to his rear beside the radio.

(3) *Switches.* The RADIO-INT switches on all control boxes, except the tank commander's, must be set on RADIO. This is the normal position for interphone operation. The tank commander's switch will be set at INT most of the time; he will change it to RADIO only as he desires radio communication. Except in an emergency, *no one but the tank commander* may operate the RADIO-INT switch on his control box. In an emergency, a member of the tank crew may communicate with the tank commander or another crew member by throwing his control box switch to INT; but this action will interrupt the tank commander's radio reception. It is the duty of the tank commander to monitor his radio receiver at all times except when speaking over the interphone or transmitting over the radio.

b. First echelon radio check. As a part of the daily Before Operation Inspection the tank commander will make the following first echelon radio check:

(1) *Cords.* (*a*) See that insulation and plugs are dry, unbroken, clean, and making good contact.

(*b*) Arrange loose cordage to prevent its entangling personnel or equipment.

(2) *Antenna.* See that—

(*a*) Mast is complete, held securely by lock screw on mast base, and sections are tight and taped.

(b) Leads at transmitter, receiver, and mast base are intact, properly insulated, and tightly connected.

(c) Mast base is clean, tight, and not cracked.

(d) Insulators passing through armor plate and bulkheads are whole and in place.

(3) *Set mountings, snaps, snubbers, etc.* Check for security and condition.

(4) *Microphones, headsets, and controls.* Check for condition and proper position. Replace from spares if necessary and turn in defective items for repair or replacement.

(5) *Spare antenna sections.* See that they are correctly placed in the roll and stowed to avoid being damaged or interfering with personnel.

(6) *Ground lead.* Check connection at both ends.

(7) *Tubes.* See that spare tubes are sealed in containers bearing date of last test. Turn in defective tubes at the earliest opportunity.

(8) *Fuses.* Check condition, and spare supply for numbers and proper rating.

(9) *Cleanliness.* See that both radio and equipment are clean.

(10) *Battery voltage.* Have driver check battery voltage. If it is low, warn cannoneer to start auxiliary generator (have this started whenever radio is operated continuously and tank engine is not running).

(11) *Crystals.* Check for number, position and frequency. Be sure required crystals are present.

c. It is the duty of each man invariably to check his personal interphone equipment upon mounting the tank; he should see that it is properly maintained, and report any difficulties to the tank commander.

d. Definite tank control, commands, and terminology are set forth in paragraph 6. The desirability and necessity of adhering to this specific language cannot be overemphasized. General conversation on the interphone causes misunderstanding and disorder and is harmful to discipline.

6. INTERPHONE LANGUAGE. a. Terms.

Tank commander	LIEUTENANT or SERGEANT
Driver	DRIVER
Gunner	GUNNER
Cannoneer	LOADER
Bow gunner	BOG
Any tank	TANK
Armored car	ARMORED CAR
Any unarmored vehicle	TRUCK
Any antitank gun	ANTITANK
Infantry	DOUGHS
Machine gun	MACHINE GUN
Airplane	PLANE

b. Commands for movement of tank.

To move forward	DRIVER MOVE OUT
To halt	DRIVER STOP
To reverse	DRIVER REVERSE
To decrease speed	DRIVER SLOW DOWN
To turn right 90°	DRIVER CLOCK 3– STEADY . . . . . ON
To turn left 60°	DRIVER CLOCK 10– STEADY . . . . . ON
To turn right (left) 180°	DRIVER CLOCK 6 RIGHT (LEFT)– STEADY . . . . . ON

To have driver move toward a terrain feature or reference point, the tank being headed in proper direction.	DRIVER MARCH ON WHITE HOUSE (HILL, DEAD TREE, ETC.)
To follow in column	DRIVER FOLLOW THAT TANK (DRIVER FOLLOW TANK NO. B-9)
To follow on road or trail	DRIVER RIGHT ON ROAD (DRIVER RIGHT ON TRAIL)
To start engine	DRIVER CRANK UP
To stop engine	DRIVER CUT ENGINE
To proceed in a specific gear	DRIVER THIRD GEAR (FIRST GEAR) (FOURTH GEAR)
To proceed at same speed	DRIVER STEADY

c. Commands for control of turret.

To traverse turret	GUNNER TRAVERSE LEFT (RIGHT)
To stop turret traverse	GUNNER STEADY ..... ON

d. Fire orders. See FM 17–12.

Section IV

CREW
DRILL

---

7. DISMOUNTED DRILL. a. To form crew. Being dismounted, the crew takes dismounted posts at the command FALL IN.

b. To break ranks. Crew being at dismounted posts, at the command FALL OUT, the crew breaks ranks. Crew members habitually fall out to the right of the tank.

c. To call off. Crew being at dismounted posts, at the command CALL OFF, the members of the crew call off in turn as follows:

(1) Tank commander _____ "SERGEANT" (or "LIEUTENANT")
(2) Gunner _____ "GUNNER"
(3) Bow gunner _____ "BOG"
(4) Driver _____ "DRIVER"
(5) Cannoneer _____ "LOADER"

d. To change designation and duties. (1) Crew being at dismounted posts, at the command FALL OUT SERGEANT (GUNNER) (DRIVER)—

(*a*) The man designated to fall out moves by the rear to the left flank position and becomes cannoneer.

(*b*) The crew members on the left of the vacated post move smartly to the right one position and prepare to call off their new designations.

(*c*) The acting tank commander starts calling off as soon as the crew is re-formed in line.

11

(2) The movement may be executed by having any member of the crew fall out except the cannoneer.

(3) All movements should be executed with snap and precision and at double time.

8. TO MOUNT. Crew being at dismounted posts.

Tank Commander	Gunner	Bow Gunner	Driver	Cannoneer
Command: PREPARE TO MOUNT.				
About face. Command: MOUNT.	About face.	About face.	About face.	About face.
Stand fast.	Mount right fender.	Stand fast.	Stand fast.	Mount left fender.
Mount right fender.	Mount right sponson.	Mount right fender.	Mount right fender.	Mount left sponson.
Mount right sponson.	Enter turret and take post.	Enter bog's seat.	Enter driver's seat.	Enter turret and take post.
Enter turret and take post.			Close battery master switches.	Turn on radio.

Connect breakaway plugs.	Connect breakaway plugs.	Connect breakaway plugs.	Connect breakaway plugs.	Connect breakaway plugs.
Command: REPORT.	Report "Gunner ready".	Report "Bog ready".	Report "Driver ready".	Report "Loader ready".

9. TO CLOSE AND OPEN HATCHES. a. To close hatches. Crew being at mounted posts.

Tank Commander	Gunner	Bow Gunner	Driver	Cannoneer
Command: CLOSE HATCHES.	Release turret traversing lock and insure that turret weapons do not block hatches.			

Tank Commander	Gunner	Bow Gunner	Driver	Cannoneer
Close hatch.		Close hatch. Raise periscope.	Close hatch. Raise periscope.	Close hatch.
Command: REPORT.	Report "Gunner ready".	Report "Bog ready".	Report "Driver ready".	Report "Loader ready".

b. To open hatches. Crew being at mounted posts.

Tank Commander	Gunner	Bow Gunner	Driver	Cannoneer
Command: OPEN HATCHES.	Release turret traversing lock and insure that turret weapons do not block hatches.	Lower periscope.	Lower periscope.	

Tank Commander	Gunner	Bow Gunner	Driver	Cannoneer
Open hatch. Command: REPORT.	Report "Gunner ready".	Open hatch. Report "Bog ready".	Open hatch. Report "Driver ready".	Open hatch. Report "Loader ready".

10. **TO DISMOUNT.** Crew being at mounted posts, turret straight ahead.

Tank Commander	Gunner	Bow Gunner	Driver	Cannoneer
Command: PREPARE TO DISMOUNT. Disconnect breakaway plugs.	Disconnect breakaway plugs.	Disconnect breakaway plugs.	Disconnect breakaway plugs. Open battery master switches.	Disconnect breakaway plugs. Turn off radio.
Command: DISMOUNT.				

Tank Commander	Gunner	Bow Gunner	Driver	Cannoneer
Emerge from turret.	Stand fast.	Emerge from hatch.	Emerge from hatch.	Emerge from turret.
Move to right sponson.	Emerge from turret.	Move to right fender.	Move to left fender.	Move to left sponson.
Move to right fender.	Move to right sponson.	Take dismounted post.	Take dismounted post.	Move to left fender.
Take dismounted post.	Move to right fender.			Take dismounted post.
	Take dismounted post.			

11. TO DISMOUNT THROUGH ESCAPE HATCH. Without weapons, crew being at mounted posts.

Tank Commander	Gunner	Bow Gunner	Driver	Cannoneer
Command: THROUGH ESCAPE HATCH, PREPARE TO DISMOUNT.				

Disconnect breakaway plugs.	Disconnect breakaway plugs. Traverse turret to give access from loader's to bog's compartment.	Disconnect breakaway plugs. Open escape hatch.	Disconnect breakaway plugs. Help bog open hatch if necessary. Open battery master switches.	Disconnect breakaway plugs. Turn radio off.
Command: DISMOUNT. Stand fast.	Stand fast.	Dismount through escape hatch.	Stand fast.	Stand fast.
	Move to left side of turret.	Crawl from under tank and take dismounted post.		Move into bog's compartment and dismount.
Move to left side of turret.	Enter bog's compartment and dismount.			Crawl from under tank and take dismounted post.

17

Tank Commander	Gunner	Bow Gunner	Driver	Cannoneer
Enter bog's compartment and dismount.	Crawl from under tank and take dismounted post.			
Crawl from under tank and take dismounted post.			Move to bog's compartment and dismount. Crawl from under tank and take dismounted post.	

12. PEP DRILL. To vary the drill routine and to keep the interest of the crew members, unexpected periods of pep drill are introduced into the training. Pep drill is a series of precision movements executed at high speed and terminating at the position of attention either mounted or dismounted. For example, the crews being dismounted, the platoon commander may command, IN FRONT OF YOUR TANKS, FALL IN; MOUNT; DISMOUNT; FALL OUT SERGEANT; ON THE LEFT OF YOUR TANKS, FALL IN; FORWARD, MARCH; BY THE RIGHT FLANK, MARCH; TO THE REAR, MARCH; MOUNT. Preparatory commands for mounting and dismounting are normally omitted from this type of drill. Posts of all crew members are changed frequently.

Section V

SERVICE OF THE PIECE

13. GENERAL. a. The crew of the howitzer consists of the gunner, who aims and fires the piece; the cannoneer, who loads the piece; and the tank commander, who controls and adjusts fire.

b. Training in service of the piece must stress rapidity and precision of movement and teamwork.

14. POSITIONS OF HOWITZER CREW. Positions of the howitzer crew are as prescribed in paragraph 4 b.

15. OPERATION OF HOWITZER. a. To open the breech. Grasp the breech operating handle and squeeze the latch until it is disengaged from its catch. Push the breech operating handle to the rear and right as far as it will go.

b. To load. Holding a round of ammunition with the right hand at the base of the cartridge case and the left hand at the middle of the assembled round, insert the nose of the projectile carefully into the chamber to avoid striking the fuze. Remove the left hand and with it grasp the operating handle. Clench the right fist, and thrust the round home into the chamber. As the rim of the cartridge case engages the extractor, it starts the closing motion of the breechblock. When this motion is felt, close the breech by moving the operating handle to the left and forward with the left hand. *Check to see that the latch locks the handle in the closed position.* Move the body and both arms to the left clear of the path of recoil, and signal "Ready" by tapping the gunner's left leg with the foot.

c. **To lay the piece.** Bring the target into the field of the telescope by the quickest practicable method, under guidance of the tank commander or by use of the periscope. To lay for direction traverse until the center line of the telescope is on the center of the target or until the proper sight picture is obtained. Make the final traversing motion against the greatest resistance, such as might be caused by cant in the tank. Then move the piece until the target shows at the proper range indicated by its relation to the range lines of the reticle. Adjustment is calculated so as to depress the muzzle with the final motion.

d. **To fire the piece.** Before firing, move the firing switch on the instrument panel to "ON". To fire, with the right heel depress the right hand firing switch button on the turret basket floor. If the piece fails to fire proceed as in paragraph 16. It may also be fired mechanically by depressing the firing pedal at the front edge of the basket floor.

e. **Safety precautions.** (1) Before loading each round, the piece will be inspected to see that there is no obstruction in the bore.

(2) The gunner must release the firing switch button of firing pedal after firing to avoid injury to the cannoneer.

(3) The gunner waits for the cannoneer's signal that the gun is loaded and he is clear of the recoil before operating the firing switch.

(4) After firing, during range and combat practice, the howitzer will be inspected by an officer to see

that it is unloaded before the tank is moved or personnel is allowed to move in front of it.

(5) In loading the piece, care must be taken not to strike the fuze or primer of a shell against any solid object; after loading, the cannoneer must take care to remain clear of the path of recoil.

(6) Stuck rounds will be removed from the bore only with rammer, cleaning and unloading M5, or with the rammer M1, which are made for this particular purpose. The method of removing is given in g and h below.

(7) Ammunition will be cleaned and inspected before stowing and each round will again be inspected before loading.

(8) Fuzes will not be disassembled or tampered with in any way.

(9) In case of a misfire, the firing switch is immediately opened before recocking. Do not touch breech mechanism until the firing switch has been opened.

(10) See safety requirements of AR 750-10.

f. **To unload an unfired round.** The cannoneer cups his hands close behind the breech to catch the base of the round as it emerges and to prevent it from slipping out and dropping to the floor. The gunner opens the breech *slowly*. (*Do not attempt to open the breech rapidly, or the case may become separated from the projectile.*) He then removes the round and returns it to its rack.

g. **To remove a stuck projectile.** If, in spite of care in opening the breech, the case and projectile do become separated, the projectile is fired out whenever possible; this is especially true in combat where unnecessary exposure of personnel is to be avoided. If it must be removed without firing the piece, the chamber should be filled with rags to form a cushion, the breech closed, and the shell rammed loose as described in **h** below and removed.

h. To unload a stuck round. When a round is stuck in the piece and it is either impossible or inadvisable to fire it out, it will be removed, except in combat, under the direct supervision of an officer. The breech being open, the cannoneer takes position to receive the round as it is pushed from the chamber, while the bow gunner or gunner dismounts and rams the round out. Using the rammer, cleaning and unloading M5, insert it in the muzzle of the gun and push it gently down the bore until it is seated on the ogive of the projectile. Exerting a steady pressure, shove the round clear so that it may be removed by the cannoneer. If the weight of several men against the staff does not suffice (*under no circumstances will the staff be used to hammer against the projectile*), apply leverage by means of a 2" x 4" piece of wood or other suitable object connected to the tank by a rope at one end, or use the rammer M1, which provides a controlled and properly cushioned blow. Keep all parts of the body as clear as possible from the muzzle or breech during the operation. If this procedure fails to remove the round, experienced ordnance personnel should be called. In combat, to avoid exposing personnel to enemy fire, the round can sometimes be pried out by using the base of an empty shell case as a lever.

16. MALFUNCTIONS. Malfunctions of the howitzer may be divided into three general classes: failure to load, failure to fire, failure to extract. Below are given the causes of the principal types of failure and the immediate action remedy to be applied.

a. Failure to load.

Failure	Cause	Immediate Action and Remedy
Round does not fully enter chamber.	Stuck round.	Remove round. Check for obstruction in chamber. Check for dirty round, and clean. Check for "bulged" (deformed) round. For removal of separated or stuck rounds see 15 g and h above.
Breech does not close.	Insufficient force in pushing round home, to clear breechblock.	Withdraw round and try again.
	Bent or undersized case rim.	Turn round so that rim engages extractors, or use new round.
	Obstruction, dirt or friction, in breech mechanism.	Remove obstruction or dirt from recess if present; otherwise remove, disassemble, clean, and lubricate breechblock.
	Worn or broken extractor.	Replace extractor.

b. Failure to fire.

Failure	Cause	Immediate Action and Remedy
Piece does not return to battery.	Obstruction between breech ring and rear portion of mount.	Drive out obstruction, or, if necessary and jack is available, use tank jack between breech ring and shoulder guard bracket of mount, to release obstruction.
	Excessive friction of tube in cradle bearing.	Relubricate. Take to ordnance if condition persists.
	Too much recoil oil.	Remove excess oil.

If piece is in battery:

Failure	Cause	Immediate Action and Remedy
Action of trigger mechanism restricted.	Safety on "Safe".	Move safety to "Fire".
Blow of firing pin fails to fire round.	Defective round.	Recock piece and attempt to fire a second time.
		Remove round to determine cause of misfire. (AR 750-10.) (See paragraph 15 for removal of live rounds.)

Failure	Cause	Immediate Action and Remedy
	Weak blow on primer due to: obstruction, dirt or friction in firing mechanism.	Dissassemble firing mechanism and remove obstruction or dirt, clean, relubricate, and assemble.
	Broken tip on firing pin.	Replace firing pin.
	Broken or weak firing spring.	Replace firing spring.
Firing pin fails to strike primer.	Obstruction, dirt, or friction in firing mechanism.	Disassemble, and remove obstruction, clean, lubricate.
	Weak or broken firing spring.	Replace.
	Defective firing pin.	Replace.
	Defective cocking lever.	Replace.
	Defective cocking fork.	Replace.
	Defective cocking lugs, on percussion mechanism.	Replace mechanism.
	Defective sear.	Replace.

c. **Failure to extract.**

Failure	Cause	Immediate Action and Remedy
Breech opens, but case is not extracted.	Broken extractor.	Pry or ram out empty case and replace extractor.
	Undersized or bent rim.	Pry or ram out.

Section VI
MOUNTED ACTION

17. TO PREPARE TO FIRE. Crew being at dismounted posts, hatches open. The antiaircraft gun is uncovered and half loaded as the tactical situation dictates.

Tank Commander	Gunner	Bow Gunner	Driver	Cannoneer
Command: PREPARE TO FIRE.				
Clean gunner's, loader's and sergeant's periscopes, gun telescope and cupola vision blocks.	Unlock traveling lock; elevate howitzer.	Lower seat. Release traveling lock.	Lower seat. Clean periscopes.	Inspect bore and chamber of howitzer.[1]
	Check traversing and elevating mechanisms.	Half load bow gun. Clean periscopes.	Close hatch; raise periscope.	Half load coaxial machine gun. Inspect smoke

[1] If tape muzzle cover is in place, inspection does not require its removal. If cover is unbroken no foreign material has entered the muzzle.

Tank Commander	Gunner	Bow Gunner	Driver	Cannoneer
	Check firing controls (including solenoids).	Check ammunition.		mortar; load mortar. Open floor compartment.
		Close hatch; raise periscope.		
Check vane sight.	Check periscope and sights.[2]			Check 105-mm rounds, smoke bombs, and machine gun ammunition.
Close hatch if desired.	Uncover and check elevation quadrant and azimuth indicator.			
Command: REPORT.	Report "Gunner ready".	Report "Bog ready".	Report "Driver ready".	Report "Loader ready".

[2] Periscope will be already raised since after the initial adjustment for the day it should not be lowered. Lowering the periscope may spoil the adjustment.

18. DUTIES IN FIRING.

Tank Commander	Gunner	Bow Gunner	Driver	Cannoneer
Give fire orders (FM 17–12). Turn on exhaust fan.	Fire on targets designated. Observe and sense all rounds through sights.	Fire on designated targets and on emergency targets that appear. When not firing, observe in assigned sector.	Turn on ventilating blower if not already operating. Observe in assigned sector and be prepared to move tank as ordered.	Load type ammunition indicated in fire order (inspect each round). Signal READY each time piece is loaded by tapping gunner on left leg. Reload all turret weapons. See that all fuzes are at DELAY unless ordered otherwise.
Observe and sense each round and notify gunner of changes in range or deflection.	Continue to fire as directed.			
Control driver with interphone.				

Tank Commander	Gunner	Bow Gunner	Driver	Cannoneer
	Call MISFIRE if piece fails to fire.			In case of misfire, check that breech is closed, piece in battery; recock piece and signal READY to gunner.
	Call STOPPAGE if coaxial gun fails to fire.			Reduce stoppages in coaxial machine gun.
	Tell loader when to fire coaxial gun if solenoid fails to operate.			Fire coaxial gun by hand when directed by gunner.
Fire AA gun.				Fire AA gun.

Determine when mortar smoke screen should be laid and give commands to produce the desired effect.	Rotate turret as directed by sergeant in adjusting smoke screen.	Keep mortar loaded at all times; adjust range, and fire immediately on command of sergeant.
When ordered by platoon commander, adjust indirect fire from forward position. Indicate aiming point to gunner.	In indirect fire: Lay piece for direction. Lay piece for elevation. Set off deflection. Fire piece on command. Make designated corrections in deflection	Keep record of ammunition expended for entry in gun book by platoon leader (number of rounds each type).

Tank Commander	Gunner	Bow Gunner	Driver	Cannoneer
	and elevation. During lulls in normal activity observe in assigned sector.			Inform sergeant when ammunition needs to be restowed. During lulls in normal activity observe in assigned sector.

19. TO SECURE GUNS[1]. In battle this operation is normally followed by RE-STOW AMMUNITION.

Tank Commander	Gunner	Bow Gunner	Driver	Cannoneer
Command: (CEASE FIRING) SECURE GUNS. Open hatch. Raise and mount	Turn off firing switch.	Clear bow machine gun;	Lower periscope.	Clear coaxial machine gun.

convoy seat.	Lock howitzer in travel position.[2]	engage traveling lock. Lower periscope.		Clear howitzer; inspect bore and close breech.
	Lock turret lock.[2]	Open hatch (first check position of howitzer).	Open hatch (first check position of howitzer).	Clear smoke mortar. Open hatch.
		Raise seat to convoy position.	Raise seat to convoy position.	
Command: REPORT.	Report "Gunner ready".	Report "Bog ready".	Report "Driver ready".	Report "Loader ready".

[1] The above drill is the minimum number of operations required to put the tank in proper condition to march after it has been prepared for combat or after range practice. If time permits, additional operations and checks are performed. The gunner checks sight adjustment and covers the elevation quadrant bubble and the azimuth indicator. The tank commander may order the bores of all weapons swabbed and their muzzles taped.

[2] Normally omitted in range procedure.

33

20. TO LOAD ALL WEAPONS. The howitzer is loaded on order. This is normally the fire order, but some types of action will dictate loading prior to the appearance of a target. Machine guns are clear until the command PREPARE TO FIRE, when they are half loaded. When the fire order is given, however, or if the unit is deployed for combat, all machine guns will be fully loaded. This does not necessarily apply to the antiaircraft gun, which is uncovered and half loaded as the tactical situation dictates.

21. USE OF AMMUNITION. a. The order of withdrawing ammunition from its stowage space in the tank is based on the principle that some readily accessible rounds always will be saved for emergency use. Other crew members will pass ammunition to the cannoneer if necessary to prevent his having to use these rounds. During combat, the position of the turret will affect the accessibility of the ammunition in various parts of the tank. In drill, however, to establish a sound method from which commanders may deviate as the need arises, the following procedure should be adhered to:

b. Ammunition is taken from its stowage space in the tank in the order: (1) Three front rows left of power tunnel; (2) racks beside bow gunner in right sponson; (3) top racks behind bow gunner. The two rear rows left of the power tunnel will be saved as a reserve for action where speed of loading is of the utmost importance. As time permits, or on the command RE-STOW AMMUNITION, rounds are moved from the racks beside the gunner in the right sponson and from the bottom racks behind the bow gunner to those which have been emptied in firing.

215

c. Upon completion of re-stowing, reports are given on the number of rounds remaining. For example the bow gunner reports, "Three smoke, six HE remaining in forward racks right sponson; one-two HE remaining right of power tunnel". The gunner reports, "Rear racks right sponson empty". The cannoneer reports, "Three smoke, three HEAT, one-nine HE remaining left of power tunnel".

22. TO LOAD AMMUNITION. Ammunition for the howitzer will be crimped upon assembly and should then be loaded and stowed with great care to avoid striking the fuze end or the primer on a hard surface, burring the rotating band, or denting the case. (See TM 9–1900.) If time is available, each crimped round should be tried in the piece before stowing to see that it can be loaded. If for some reason rounds cannot be crimped, each case should be tried in the piece prior to assembly of the round. All rounds of HE will be set at FUZE DELAY at this time. Both howitzer and machine gun ammunition will be passed through the hatches as most convenient under the circumstances, a man being stationed on the forward or rear hull to relay it to those in the tank.

Section VII
DISMOUNTED ACTION

23. TO FIGHT ON FOOT. a. Crew being at mounted posts, hatches open. Crew members, including the tank commander, keep below hatches until completely ready to dismount and go into action and until the order DISMOUNT is given.

Tank Commander	Gunner	Bow Gunner	Driver	Cannoneer
Command: PREPARE TO FIGHT ON FOOT.				
Disconnect breakaway plugs.	Disconnect breakaway plugs.	Disconnect breakaway plugs.	Disconnect breakaway plugs.	Disconnect breakaway plugs.
Order distribution of grenades.		Pass tripod to driver.	Receive tripod from bog.	

Take hand grenades, submachine gun and 6 clips ammunition.	Procure grenades as ordered.	Procure grenades as ordered.	Procure grenades as ordered.	Procure grenades as ordered.
	Stand fast.			
		Install elevating mechanism on bow gun; dismount gun; install pintle.	Procure 3 boxes cal .30 ammunition.	Help driver get ammunition. Take 1 box cal .30 ammunition.
Stand fast.		Pass submachine gun and 6 clips ammunition to driver.	Receive bog's submachine gun and ammunition.	Stand fast.
		Take spare parts roll and spare bolt assembly.		

Command: DISMOUNT.

37

Tank Commander	Gunner	Bow Gunner	Driver	Cannoneer
Dismount via right sponson and fender.				Dismount to left sponson.
				Receive tripod from driver.
	Dismount to right sponson. Receive bow gun from bog.		Pass tripod to loader.	
Receive 2 boxes cal .30 ammunition from driver.		Pass bow machine gun to gunner.	Pass 2 boxes cal .30 ammunition to sergeant.	Dismount. Set up tripod.
Cover dismounting of crew.	Dismount. Mount bow gun; man gun as No. 2.	Dismount. Receive box cal .30 ammunition and submachine gun and ammunition from driver.	Pass box cal .30 ammunition and submachine gun and ammunition to bog.	Help mount bow gun; man gun as No. 1.
Act as squad leader of machine gun squad.			Move into turret; connect	

38

breakaway plugs; maintain contact with platoon leader.

Man gun as No. 3.

b. The dismounted crew moves to the position indicated by the tank commander or, in drill, 5 yards in front of the tank. The crew members take the posts and perform the duties of the crew of a ground-mounted machine gun as prescribed for gun drill in FM 23-55 (1944 edition).

c. In combat it is assumed that the tank will be moved to a concealed position if possible, before the crew dismounts. Otherwise the driver will move the tank to a concealed position before mounting to the turret.

24. TO MOUNT FROM DISMOUNTED ACTION.

Tank Commander	Gunner	Bow Gunner	Driver	Cannoneer
Command: OUT OF ACTION.				
Supervise taking gun out of action.	Dismount machine gun.	Take mounted post (leave cal .30 ammu-	Disconnect breakaway plugs.	Help dismount machine gun.

39

Tank Commander	Gunner	Bow Gunner	Driver	Cannoneer
Cover other crew members with submachine gun. Pass remaining cal .30 ammunition to driver.	Pass bow machine gun to bog.	nition and submachine gun and ammunition in front of tank).	Resume mounted post. Receive tripod; place in bog's compartment.	Fold tripod. Pass tripod to driver.
Pass bog's submachine gun and ammunition to him.	Take mounted post.	Receive and mount bow gun (remove and stow ground accessories).	Receive remaining ammunition; place near loader.	Mount tank with remainder of box cal .30 ammunition.
Take mounted post.	Receive and stow grenades.	Receive and stow submachine gun and ammunition.	Connect breakaway plugs.	Take mounted post. Stow ammunition.
Return grenades. Stow submachine gun and ammuni-		Return grenades.		Receive and stow grenades.

tion. Connect breakaway plugs.	Connect breakaway plugs.	Stow spare parts roll and spare bolt assembly. Stow tripod. Connect breakaway plugs.		Connect breakaway plugs.
Command: REPORT.	Report "Gunner ready".	Report "Bog ready".	Report "Driver ready".	Report "Loader ready".

25. TO ABANDON TANK. If it becomes necessary to abandon tank, the crew proceeds as in paragraph 10 or 11 with the following changes or additions:

a. Time permiting deliberate action, the tank commander displays the flag signal DISREGARD MY MOVEMENTS, and supervises the disabling of those weapons which remain in the tank. Backplates are removed from machine guns and the firing pin and guide from the howitzer. All similar spare parts are also removed. Individual weapons and maximum possible ammunition loads are carried. The driver dismounts in order with the rest of the crew.

41

b. Ordinarily the tank is abandoned as a result of a direct hit either causes it to catch fire or disables it so that it becomes a vulnerable target. In such instances there may be less than five seconds in which the crew can escape without further injury. At the command ABANDON TANK, crew members throw open hatches, climb out, jump to ground and take cover at a safe distance from the tank. It is particularly important in case of fire to hold the breath until clear of the vehicle. Inhaling the fumes and smoke of the fire may injure the lungs and will at least incapacitate the individual for a time.

26. TO DESTROY TANK. When the command DESTROY TANK is given, crew members first remove what equipment is to be carried away. They then destroy the tank, weapons, ammunition, and equipment to be left, as prescribed in Section XI.

27. ACTION IN CASE OF FIRE. a. Fire in engine compartment. The first crew member to discover fire calls, ENGINE FIRE.

Tank Commander	Gunner	Bow Gunner	Driver	Cannoneer
Disconnect breakaway plugs.	Disconnect breakaway plugs.	Disconnect breakaway plugs.	Disconnect breakaway plugs.	Disconnect breakaway plugs.
Dismount to rear deck.		Take hand extinguisher.	Pull ONE fixed extinguisher	Obtain wrenches.

Receive wrenches and fire extinguisher.	Pass hand extinguisher to sergeant.		control handle; shut off engine.	Pass wrenches to sergeant.
	Dismount to rear deck.	Dismount.	Dismount.	Dismount.
Start to open top engine doors.		Go to rear of tank; unfasten rear engine doors, ready to open if needed.	Go to rear of tank and help as ordered.	Go to rear of tank and assist other crew members.
If fixed extinguisher has not put out fire, use hand extinguisher through top doors or order use of second fixed extinguisher.	Unfasten top engine doors. Stand by to pull exterior control handle of second fixed extinguisher if ordered.	Use hand extinguisher through rear doors if ordered.		

43

b. Fire in air horn. (Applicable only to tanks equipped with radial engines.) The first crew member to discover fire calls, AIR HORN FIRE.

Tank Commander	Gunner	Bow Gunner	Driver	Cannoneer
Disconnect breakaway plugs.	Disconnect breakaway plugs.	Disconnect breakaway plugs.	Disconnect breakaway plugs.	Disconnect breakaway plugs.
Take wrench and screwdriver from loader; dismount.	Take hand extinguisher. Dismount.	Take hand extinguisher; dismount.	Race engine (if cranking, continue in attempt to start).	Obtain wrench and screwdriver; pass to sergeant.
Go to rear of tank; open rear engine doors.	Remove cone from hand extinguisher nozzle.	Remove cone from hand extinguisher nozzle.	Dismount if ordered.	Dismount to rear deck.
If racing engine has not put out fire, cut small hole with screwdriver in air horn-intake tube coupling.	Go to rear of tank; stand by to use extinguisher.	Go to rear of tank. Insert extinguisher nozzle in hole made by sergeant; operate extinguisher.		Stand by to operate fixed extinguisher if ordered.

c. Fire in fighting compartment. The first crew member to discover the fire calls, TURRET (or HULL) FIRE. The tank is stopped and the engine shut off. Fire extinguishers are passed to the men nearest the fire, and the crew members nearest them help in any way possible to extinguish the fire. The turret is traversed if necessary. The tank commander supervises the work and orders the crew to dismount if the fire gets beyond control.

28. ADVICE TO INSTRUCTORS. a. Disciplined and effective dismounted action requires long and arduous drill. Satisfactory results can be obtained only by painstaking repetition of each movement. The technique of mounting and dismounting of all crew members is observed in detail by the tank and platoon commanders and altered, if necessary, before habits are formed. Once each man has found the most efficient method of mounting and dismounting, he is encouraged to adhere rigidly to it.

b. Training in dismounted action is best undertaken in the field rather than in the tank park. Crews are required to dismount to fight on foot on all types of terrain, and under every variety of simulated combat conditions, with full loads of ammunition. Rough terrain complicates the problem of dismounting through the escape hatch, and develops ingenuity and physical agility not possible in tank park training.

c. Instructors must explain and demonstrate to tank crews how necessary to their safety and success in combat is a high state of training in dismounted action. They must point out that skill and practice in use of the escape hatch will pay dividends. The crew keeps the escape hatch door clean and well lubricated so that its release is immediate and positive. Frequent inspection of the mechanism is made by the tank commander to see that the locking rods are not bent.

29. GENERAL PRECAUTIONS. a. Fire prevention.
(1) Smoking in or on the tank is prohibited.

(2) During fueling a crew member stands on the rear deck holding a fire extinguisher with the nozzle trained on the fuel inlet, ready to use it instantly if needed.

(3) Use of gasoline for cleaning any part of the tank is prohibited.

b. Mounting and operating tank. (1) Crew members mount and dismount by the front of the tank except during range practice.

(2) Unnecessary contact with any part of the weapons or sighting equipment will be avoided. This includes—

(*a*) Stepping on the howitzer barrel or shield, or the machine guns in mounting or dismounting.

(*b*) Supporting oneself by holding the tube, howitzer shield, or machine guns in mounting or dismounting.

(*c*) Use of the shoulder guard as a step in entering or leaving the turret.

(3) Crash helmets if available, or helmet liners are worn at all times inside the tank.

(4) In operating cross country the tank commander warns the driver and crew when the tank approaches rough terrain.

(5) Where possible the driver avoids rough or uneven ground which might cause injury to the tank or crew.

(6) In traveling with hatches open over rough ground or through woods, crew members constantly

check the engagement of the cover latching mechanism and the security of covers in the open position.

(7) The antenna is lowered to prevent contact with low branches or low-hanging wires, especially those which may carry high voltage electricity.

(8) The tank is driven in low range when being moved forward in confined spaces.

c. Park and bivouac precautions. (1) Sleeping underneath, behind, or in front of tanks should be prohibited.

(2) In moving a tank in park or bivouac—

(*a*) A guide is always employed to direct the movement.

(*b*) The guide's position is at least ten feet in front of the tank and to one side, clear of its path, in directing the tank either forward or back.

(*c*) At night the guide is especially charged with seeing that the path ahead of and behind the tank is clear of personnel, particularly those sleeping on the ground.

(*d*) The guide moves at a walk to avoid stumbling on uneven ground.

d. Miscellaneous. (1) After machine guns are cleared a cleaning rod is pushed through the barrel and chamber to insure that the chamber is empty. A T-block is then inserted into the receiver.

(2) Tank weapons, except the antiaircraft gun, are fired only when the driver's and bow gunner's hatches are closed.

(3) Care will be taken, while working about a running engine to keep fingers and hands away from fans; fan belts, drive shafts, and other moving parts.

(4) 105-mm ammunition will be securely stowed.

(5) Ammunition will not be carried on the rear deck.

(6) No items of equipment will be carried on the rear deck in such a manner as to block the air inlet grilles.

(7) There is danger of monoxide poisoning for the crew of a towed tank when the medium tank or a tank recovery vehicle mounted on tank chassis is used as the towing vehicle. This danger is greatest when the towing vehicle is powered with a radial engine, and when a short hitch, such as that obtained with the towing bar, is used. Men should be kept out of the towed tank wherever possible; but where this is not possible, frequent periodic check of the occupants of the towed vehicle should be made.

Section VIII

EVACUATION OF WOUNDED FROM TANKS

30. GENERAL. Wounded members of the tank crew will normally be removed from disabled tanks by their fellow crew members. The operation requires the utmost speed to save the lives of those who are unhurt as well as of the casualty. A tank set afire by an enemy hit can trap its crew in a matter of seconds; and an enemy who has determined the range and disabled a tank with a direct hit will probably continue shooting until the vehicle burns. It is essential, therefore, that all crew members become extremely proficient in the quickest methods of removing one another from the tank. Speed is the primary requisite; care in handling will be stressed only where it has been possible to move the tank to cover. If the action has ceased momentarily, or the tank has been able to disengage itself without hindering the accomplishment of the mission, the casualty is removed on the spot and then carried to a protected place where emergency first aid is administered. Otherwise the action will be continued until such an opportunity is presented.

31. METHODS EMPLOYED. The methods of evacuation described herewith are based on a two-man team, which is the largest number than can effectively work around a single hatch opening. In some cases a third man will be able to give considerable help from inside by placing belts around the wounded man or by moving him to a position where he can be grasped from above. Speed will usually dictate that the casualty be grasped by portions of his clothing or by the arms for removal. If an arm is broken, however, or if there are other injuries which will be aggravated by such procedures and if time allows, some form of sling may be improvised which will relieve the part from further injury. Only equipment which is immediately available, like pistol belts, web belts, or field bag straps, will be used for this purpose. Suggested uses of some of these items, as well as more elaborate techniques of evacuation, will be found in FM 17–80.

32. DRILL. This paragraph suggests two drills which may be used as models for evacuating crew members from any position. The composition of the evacuating team should be changed frequently to provide practice for all members of the crew in meeting various emergencies.

a. The first member of the crew to discover that another is hit and so badly wounded as to require his removal calls, BOG (LOADER) (SERGEANT) WOUNDED. If the tank is not then actively engaged and the tank commander decides that evacuation is necessary, he commands, EVACUATE BOG. The other crew members dismount, unless one man is needed to help from inside; and the two nearest the hatch above the wounded man go to that hatch to act as the evacuation crew. If the man nearest the casualty in the tank sees that his help is needed, he stays inside and immediately starts to arrange a sling or take whatever other steps will speed the operation. One of the crew takes the first aid kit with him in dismounting, or it is removed at the first opportunity thereafter. The remaining crew member, if available, helps in lowering the casualty to the ground. Before leaving the wounded man, whose position is marked so that he will not be run over, the tank commander reports by radio that he has lost one or more men and gives the location where they may be found.

b. To evacuate Bog (Driver). Tank commander commands, EVACUATE BOG. Driver or gunner unlocks bow gunner's hatch from inside; No. 2 opens hatch from outside.

No. 1	No. 2
Kneel on inner edge of hatch.	Take position to the outside rear of hatch.
Reach into hatch and grasp hands of casualty, straightening him in seat if necessary.	
Cross arms over chest.	Grasp nearest hand when arms are crossed.
Raise and rotate casualty so that he faces outward.	Raise casualty and help rotate him outward.
Seat casualty on front rim of hatch; support in this position while No. 2 dismounts.	Help seat casualty; dismount to ground in front of Bog's hatch.
Lower trunk into arms of No. 2.	Receive and support trunk of wounded man, holding it beneath arms around chest.
Lift legs out of hatch as No. 2 lowers body along slope plate.	Lower body along slope plate and support until No. 1 can reach ground and assist.
Dismount. Place casualty in carry position.	Place casualty in carry position.
Carry casualty to protected area.	Help No. 1 carry to protected area.

232

c. To evacuate cannoneer[1]. Tank commander commands, EVACUATE LOADER. He dismounts to rear deck to act as No. 1. Gunner stays in the turret to act as No. 2. If time permits he traverses the turret until the hatch is near the center of the rear deck.

No. 1	No. 2
Take position on rear deck behind turret hatch.	Raise casualty as high as possible in hatch opening, holding around chest.
Grasp casualty by hands.	
Raise casualty through hatch and seat on rear edge.	Help No. 1 raise casualty by lifting from below.
Hold casualty while No. 2 dismounts to rear deck.	Dismount to rear deck.
Pick casualty up in arms; carry to rear and lay along back edge of deck.	Help No. 1 pick up casualty and carry to rear of tank; dismount.
Help No. 2 lift trunk of casualty off tank; dismount.	Lift upper part of body off tank and support until No. 1 arrives to help.
Lift hips and legs off tank.	
Carry casualty to protected area.	Help carry casualty to protected area.

[1] Drill applicable in this form only where the casualty can be lifted by his arms, especially in the case of a big man whose shoulders are too wide for the hatch opening when his arms are lowered. In such cases the cannoneer is evacuated through the cupola hatch.

42. AFTER OPERATION MAINTENANCE, M4A3. a. After operation the tank is immediately given whatever servicing and maintenance is needed to prepare it in every way for further sustained action. This servicing covers all the points listed in the Before Operation Inspection and covers them in the same order, with obvious modifications. (For example, the tank is locked at the end of the inspection instead of being unlocked at the beginning; the check for leaks under the tank is more effective after it has stood for awhile; battery switches are turned off rather than on and only after all checks requiring use of battery power; equipment is covered and stowed rather than being uncovered and made ready for use.)

b. The tank will be completely cleaned, serviced, and replenished (fuel, oil (all types), grease, coolant, ammunition (all types), first aid kit, water, and rations). *All special precautions against fire will be observed while refueling.* Crew members will perform the following additional operations not covered in the Before Operation Inspection.

Tank Commander	Gunner	Bow Gunner	Driver	Cannoneer
Command: PERFORM AFTER OPERATION MAINTENANCE.				
Complete trip ticket; forward to	Clean all weapons.		Idle engine 2 minutes be-	Help gunner clean

platoon leader, together with report of any necessary 2d echelon maintenance, fuel, lubricants, ammunition and rations required.

Help driver clean tank.

fore stopping. Clean tank suspension and outside of tank.

weapons.

Help gunner clean weapons.

Help gunner clean weapons.

43. PERIODIC ADDITIONAL SERVICES, M4A3. Services performed weekly in garrison; in combat and on maneuvers they are performed after each field operation.

Tank Commander	Gunner	Bow Gunner	Driver	Cannoneer
Command: FALL IN; PREPARE FOR INSPECTION.				
Inspect crew.	Stand inspection.	Stand inspection.	Stand inspection.	Stand inspection.
Command: PERFORM INSPECTION.				

Tank Commander	Gunner	Bow Gunner	Driver	Cannoneer
Supervise inspection.	Mount to turret. Clean turret. Clean and touch up any rust spots in turret.	Mount to rear deck; help driver clean engine and engine compartment.	Open engine doors and clean engine and engine compartment. Take mounted post.	Mount to turret. Clean batteries and case. Test batteries with hydrometer. Bring cells to proper water level.
	Dismount.	Operate and check hull drain valves in engine compartment. Take mounted post.	Clean driver's compartment and left interior of hull.	Operate auxiliary generator to charge batteries. Dismount.

	Clean bog's compartment and right interior of hull.	Operate and check hull drain valve.	
	Operate and check hull drain valve.		
		Drive tank forward as required for tightening wedge nuts.	Help gunner tighten wedge nuts and inspect track.
Tighten all wedge nuts and inspect track.		Perform 250-mile lubrication, referring to appropriate guide.	
Help perform 250-mile lubrication.	Help perform 250-mile lubrication.	Close engine doors.	Help perform 250-mile lubrication.

Tank Commander	Gunner	Bow Gunner	Driver	Cannoneer
Command: REPORT.		Take mounted post; clean and touch up rust spots in bog's compartment.	Take mounted post; clean and touch up rust spots in driver's compartment.	
	Report "Gunner ready".	Report "Bog ready".	Report "Driver ready".	Report "Loader ready".

96

Section X

DESTRUCTION OF EQUIPMENT

44. GENERAL. a. The destruction of materiel requires a command decision and will be undertaken only on authority delegated by division or higher commanders. Destruction is ordered only after every possible measure for the preservation or salvage of the materiel has been taken, and when in the judgment of the person exercising the authority such action is necessary to prevent.

(1) Its abandonment in the combat zone.

(2) Its capture intact by the enemy.

(3) Its use by the enemy, if captured, against our own or allied troops.

(4) Knowledge of its existence, functioning, or exact specifications from reaching enemy intelligence.

b. The principles to be followed are—

(1) Methods for the destruction of materiel subject to capture or abandonment in the combat zone must be adequate, uniform, and easily followed in the field.

(2) Destruction must be as complete as available time, equipment, and personnel will permit. If thorough destruction of all parts cannot be completed, the most important features of the materiel should be destroyed, and parts essential to the operation or use of the materiel which cannot be easily duplicated, should be ruined or destroyed. The same essential parts must be destroyed on all like units to

prevent the enemy from constructing one complete unit from several damaged ones by "cannibalism".

c. Crews will be trained in the prescribed methods of destruction, but training will not involve the actual destruction of materiel.

d. (1) The methods outlined in the paragraphs below are given in order of effectiveness. If method No. 1 cannot be used, destruction should be accomplished by one of the other methods in order of priority shown. Adhere to the sequences.

(2) Certain methods require special tools and equipment, such as TNT and incendiary grenades, which normally may not be items of issue. The issue of such special tools and material, the vehicles for which issued, and the conditions under which destruction will be effected are command decisions in each case, according to the tactical situation.

45. DESTRUCTION OF HOWITZER. Remove sights. If evacuation is possible, carry the sights; if evacuation is not possible, thoroughly smash all periscopic sights and the telescope.

a. **Method No. 1.** (1) Open drain plugs on recoil mechanism, allowing recoil fluid to drain. *It is not necessary to wait for the recoil fluid to drain completely before firing the howitzer as in (4) below.*

(2) Place an *armed (safety pin removed)* antitank grenade, HE, or *armed (safety pin removed)* antitank rocket in the tube about 6 inches in front of, and with the ogive nose end toward, the HE shell in (3) below.

(3) Set fuze on an HE shell at "superquick", insert shell in the piece and close the breech.

(4) Attach a piece of string to the howitzer firing linkage in such a way that it may be fired by pulling

the string. Dismount from the tank (down to the left rear) and fire the piece. Elapsed time: Approximately 2 to 3 minutes.

b. Method No. 2. Insert from three to five ½-pound TNT blocks in the bore near the muzzle, eight to ten in the chamber. Close the breechblock as far as possible without damaging the safety fuze. Plug the muzzle tightly with earth to a distance of approximately 11 inches from the muzzle. Detonate the TNT charges simultaneously.

c. Method No. 3. With another gun, fire HE, HEAT or AP projectiles at the tube of the piece until it is rendered useless.

d. Method No. 4. Insert four unfuzed M14 incendiary grenades, end to end, midway in the tube at 0° elevation. Ignite these four grenades with a fifth equipped with a 15-second Bickford fuze. The metal from the grenades will fuse with the tube and fill the grooves. Elapsed time: 2 to 3 minutes.

46. DESTRUCTION OF MACHINE GUNS. a. Method No. 1. (1) *Caliber .30 machine gun.* Field strip. Use barrel as a sledge. Raise cover until vertical; smash cover down toward front. Deform and break backplate; deform T-slot. Wedge lock frame, back down, into top of casing between top plate and extractor cam; place chamber end of barrel over lock frame depressors and break off depressors. Insert barrel extension into back of casing, allowing the shank to protrude; knock off shank by striking with barrel from the side. Deform and crack casing by striking with barrel at side plate corners nearest feedway. Elapsed time: 2½ minutes.

(2) *Caliber .50 machine gun.* Field strip. Use barrel as a sledge. Raise cover; lay bolt in feedway; lower

cover on bolt; smash cover down over bolt. Deform backplate. Wedge buffer into rear of casing allowing depressors to protrude; break off depressors by striking with barrel. Lay barrel extension on its side. Hold down with one foot, break off the shank. Deform casing by striking side plates just back of the feedway. Elapsed time: 3½ minutes.

b. Method No. 2. Insert bullet point of complete round into muzzle and bend case slightly, distending mouth of case to permit pulling of bullet. Spill powder from case, retaining sufficient powder to cover the bottom of case to a depth of approximately ⅛ inch. Re-insert pulled bullet, point first, into the case mouth. Chamber and fire this round with the reduced charge; the bullet will stick in the bore. Chamber one complete round, lay weapon on ground, and fire with a 30-foot lanyard. Use the best available cover, as this means of destruction may be dangerous to the person destroying the weapon. Elapsed time: 2 to 3 minutes.

c. Small arms. Small arms cannot be adequately destroyed by firing with the bore stuck in the ground, with or without a bullet jammed in the muzzle.

d. Machine gun tripod mount, caliber .30 M2. Use machine gun barrel as a sledge. Deform traversing dial. Fold rear legs, turn mount over on head, stand on folded rear legs, knock off traversing dial locking screw, pintle lock, and deform head assembly. Deform folded rear legs so as to prevent unfolding. Extend elevating screw and bend screw by striking with barrel; bend pintle yoke. Elapsed time: 2 minutes.

47. DESTRUCTION OF TANK. a. Method No. 1. (1) Remove and empty the portable fire extinguishers.

Smash the radio (paragraph 52). Puncture fuel tanks. Use fire of caliber .50 machine gun, or a cannon, or use a fragmentation grenade for this purpose. Place TNT charges as follows: 3 pounds between engine oil cooler and right fuel tank; 2 pounds under left side of transmission as far forward as possible. Insert tetryl nonelectric caps with at least 5 feet of safety fuse in each charge. Ignite the fuses and take cover. Elapsed time: 1 to 2 minutes, if charges are prepared beforehand and carried in the vehicle.

(2) If sufficient time and materials are available, additional destruction of track-laying vehicles may be accomplished by placing a 2-pound TNT charge about the center of each track-laying assembly. Detonate those charges in the same manner as the others.

(3) If charges are prepared beforehand and carried in the vehicle, keep the caps and fuses separated from the charges until used.

b. Method No. 2. Remove and empty the portable fire extinguishers. Smash the radio (paragraph 52). Puncture fuel tanks (see *a* (1) above). Fire on the vehicle using adjacent tanks, antitank or other artillery, or antitank rockets or grenades. Aim at engine, suspension, and armament in the order named. If a good fire is started, the vehicle may be considered destroyed. Elapsed time: About 5 minutes per vehicle. Destroy the last remaining vehicle by the best means available.

48. DESTRUCTION OF AMMUNITION. a. General. (1) Time will not usually permit the destruction of all ammunition in forward combat zones.

(2) When sufficient time and materials are available, ammunition may be destroyed as indicated

below. At least 30 to 60 minutes may be required to destroy adequately the ammunition carried by combat units.

(3) In general, the methods and safety precautions outlined in Chapter 4, TM 9–1900, should be followed whenever possible.

b. Unpacked complete round ammunition. (1) Stack ammunition in small piles. (Small arms ammunition may be heaped.) Stack or pile most of the available gasoline in cans and drums around the ammunition. Place on pile all available inflammable material such as rags, scrap wood, and brush. Pour the remaining available gasoline over the pile. Sufficient inflammable material must be used to insure a very hot fire. Ignite the gasoline and take cover.

(2) Destroy 105-mm ammunition by sympathetic detonation, using TNT. Stack the ammunition in two stacks about 3 inches apart, with fuses in each stack toward each other. Place TNT charges between the stacks. Use 1 pound of TNT per four or five rounds of ammunition. Detonate all charges of TNT simultaneously from cover.

c. Packed complete round ammunition. (1) Stack the boxed or bundled ammunition in small piles. Cover with all available inflammable materials, such as rags, scrap wood, brush, and gasoline in drums or cans. Pour gasoline over the pile. Ignite the gasoline and take cover. (Small arms ammunition must be broken out of the boxes or cartons before burning.)

(2) (*a*) The destruction of packed complete round ammunition by sympathetic detonation with TNT is not advocated for use in forward combat zones. To insure satisfactory destruction involves putting TNT in alternate cases or bundles of ammunition, a time-consuming job.

(*b*) In rear areas or fixed installations, sympathetic detonation may be used to destroy large ammunition supplies if destruction by burning is not feasible. Stack the boxes, placing in alternate boxes in each row sufficient TNT blocks to insure the use of 1 pound of TNT per four to five rounds of 105-mm ammunition. Place the TNT blocks at the fuse end of the rounds. Detonate all TNT charges simultaneously. See FM 5–25 for details of demolition planning and procedure.

d. Miscellaneous. Grenades, antitank mines, and antitank rockets may be destroyed by the methods outlined in b and c above for complete rounds. The amount of TNT necessary to detonate these munitions is considered less than that required for detonating artillery shells. Fuses, boosters, detonators, and similar material should be destroyed by burning.

49. FIRE CONTROL EQUIPMENT. Fire control equipment, including optical sights and binoculars, is difficult to replace. It should be the last equipment to be destroyed. If evacuation of personnel is made, all possible items of fire control equipment should be carried. If evacuation of personnel is not possible, fire control equipment must be thoroughly destroyed as indicated below.

a. Firing tables, trajectory charts, slide rules and similar items should be thoroughly burned.

b. All optical equipment that cannot be evacuated will be thoroughly smashed.

50. RADIO EQUIPMENT. a. Books and papers. Instruction books, circuit and wiring diagrams, records of all kinds for radio equipment, code books, and registered documents will be destroyed by burning.

b. Radio sets. (1) Shear off all panel knobs, dials, etc., with an ax. Break open the set compartment by smashing in the panel face, then knock off the top, bottom, and sides. The object is to destroy the panel and expose the chassis. On top of the chassis, strike all tubes and circuit elements with the ax head. On the under side of the chassis, if it can be reached, use the ax to shear or tear off wires and small circuit units. Break sockets and cut unit and circuit wires. Smash or cut tubes, coils, crystal holders, microphones, earphones, and batteries. Break mast sections and break mast base at the insulator.

(2) When possible, pile up smashed equipment, pour on gas or oil, nad set it on fire. If other inflammable material, such as wood, is available, use it to increase the fire. Bury smashed parts.

INDEX

	Paragraphs	Pages
Abandon tank	25	41
Action:		
Dismounted (Section VII)	23-29	36
Mounted (Section VI)	17-22	27
Advice to instructors	28	45
After operation maintenance	37, 42	70, 92
Ammunition:		
Destruction	48	101
Expenditure record	18	31
Expenditure report	18	32
Handling in tank	15, 18	19, 31
Inspection before firing	15, 16, 17	20, 22, 27
Loading tank	22	35
Report of rounds remaining	21	35
To re-stow	21	34
Use	21	34
Auxiliary generator:		
Operate	38	73
Start	34, 39	58, 80
Book, gun	34, 39	63, 85
Breech:		
Close	15	19
Open	15	19
Crew:		
Composition	3	2
Control (Section III)	5, 6	4, 9
Drill (Section IV)	7-12	11
Formation	4	2
To dismount	10, 11	15, 16
To form	7	11
To mount	8	12
Destroy tank	26	42
Destruction of equipment (Section XI)	44-50	97
Fire control equipment	49	103

	Paragraphs	*Pages*
Destruction of equipment (cont)		
Tank	47	100
Weapons	45, 46	98, 99
Disable weapons	25	41
Dismounted action		
(Section VII)	23-29	36
To mount from	24	39
Dismounting:		
Tank crew	10	15
Through escape hatch	11	16
Training in the field	28	45
Driver's report	33	53
Duties in firing	18	29
Escape hatch:		
Dismount through	11	16
Inspection of	28	45
Evacuation of wounded from		
tanks (Section VIII)	30-32	49
Expenditure of ammunition	18	31
Fire, action in case of	27	42
Fire control equipment,		
destruction	49	103
Fire extinguishers	27	42
Checking fixed (Phase B)	34, 39	60, 80
Checking hand (Phase C)	34, 39	61, 82
Fire, prepare to	17	27
Firing:		
Duties of crew in	18	29
Switch trigger	15	20
First echelon radio check	5	7
Formations	4	2
Gun book	34, 39	63, 85
Hatches:		
Close and open	9	13
Dismount through escape	11	16
Howitzer crew:		
Composition	13	19
Positions	4	2

	Paragraphs	*Pages*
Howitzer, tank:		
Destruction	45	98
Firing	15	19
Inspections (Section IX)	33-43	53
Laying	15	20
Loading	15	19
Operation	15	19
Remedies for malfunctions	16	22
Unloading	15	21
Inspections and maintenance (Section IX)	33-43	53
After operation maintenance	37, 42	70, 92
Before operation	34, 39	54, 75
During operation	35, 40	66, 87
Halt	36, 41	66, 88
Periodic additional services	38, 43	72, 93
Radio	5	7
Vehicle	33-43	53
Weapons	15, 33-43	22, 53
Interphone:		
Language	6	9
Operation	5	4
Laying gun	15	20
Load all weapons	20	34
Loading:		
Ammunition in tank	22	35
Half (on command PREPARE TO FIRE)	20	34
Howitzer	15	19
Maintenance after operation	37	70
Monoxide poisoning in towed tanks	29	48
Mounted action (Section VI)	17-22	27
Mounting tank crew	8	12
Muzzle covers	17, 19, 34, 39	27, 33, 59, 79
Operation of howitzer	15	19
Operation of interphone and radio	5	4

	Paragraphs	*Pages*
Pep drill	12	18
Posts:		
Dismounted	4	2
Mounted	4	2
Prepare to fire	17	27
Radio:		
Destruction	50	103
Inspection	5, 34, 39	7, 64, 85
Operation	5	4
Remedies for malfunctions	16	22
Report of ammunition remaining	21	35
Re-stow ammunition	21	34
Safety precautions	15, 29	20, 46
Secure guns	19	32
Service of the piece (Section V)	13-16	19
Stoppages	16, 18	22, 30
Stuck round	15, 16	22
Tank:		
Control	5, 6	4, 9
Destruction	47	100
Technique of dismounting	10, 11, 28	15, 16, 45
Unload:		
A stuck round	15	22
The howitzer	15	21
Weapons:		
Destruction	45, 46	98, 99
Inspection	15, 34	20, 54
Half load	17, 20	27, 34
Load	15, 18, 20	27, 29, 34
Unload	15, 19	21, 32

CROMWELL MARK I
SERVICE INSTRUCTION BOOK

First Edition – 1943

251

INTRODUCTION

The Cromwell I is a heavy-weight fighting vehicle. The front part of the hull accommodates the crew of five, while the rear part houses the engine and transmission. The front part is further subdivided into a fighting compartment, a driver's compartment and a forward gunner's compartment. A partition, with access holes cut in it, separates the personnel in the fighting compartment from the driver and front gunner, while a similar partition, also with an access hole, separates the driver from the front gunner.

Single armour plate is used at the front and rear of the vehicle, but a double plate is placed at the sides, the outer plate of which affords protection to the suspension. The inner plate, on each side, carries five main suspension spring housings, each of which is attached to an axle arm. These arms are pivoted in bushes housed in the main suspension tubes which are attached to the floor plates.

The axle arms project through the outer side plates to carry five road wheels on each side of the vehicle, which run on tracks. The tracks are driven by double-toothed sprocket wheels at the rear of the vehicle and have tensioner wheels at the front.

The turret consists of double armour plated sides, the outer plates of which are bolted to the welded inner structure, and a single plate roof. It is rotated hydraulically.

Three main cooling air intakes are fitted, one at each side of the engine compartment roof and another on top of the roof. The air outlet louvres, of which there are three, are at the rear.

The hull is sealed against water ingress to a depth of 4 ft., by means of a fording flap in the lower air outlet louvre.

There are four external access doors, one to the driver's compartment, two in the turret roof and one to the front gunner's compartment, which can also be used as an emergency escape door.

For fire fighting, $C.O._2$ carbon dioxide apparatus is provided.

A 12-cylinder V-type engine is installed, which is connected to a dry twin-plate clutch. From this clutch the power is transmitted through a five-speed gearbox to the final drives on which the driving sprockets are mounted.

The main oil and fuel tanks are located on each side of the engine, while the coolant radiators are installed at the rear of the engine compartment. Steering is accomplished through controlled epicyclics and the turning circle is dependent upon the gear ratio selected by the driver. The clutch, steering and track brakes are all hydraulically operated.

It is important to remember that the vehicle will turn about its axis with the gear in neutral (see Fig. 5).

The vehicle has been designed so that the crew can be as comfortable as practicable, and yet at the same time all available space has been used to the maximum advantage. In this way, the vehicle presents the smallest possible target for its fire power and weight.

VEHICLE SPECIFICATION

Weight	27 tons (approx.) fully stowed, including crew.
Power weight ratio	22 to 1 b.h.p./tons.
Overall length	20 ft. 10 in.
Overall width	9 ft. 6½ in. (over stiffener grooves).
Overall height	8 ft. 2 in. (maximum over aerial mounting).
Crew	Total 5 (commander, two gunners, loader and driver).
Armament	One 6-pdr. One 7·92 mm. Besa M.G. } co-axially mounted in turret. One 7·92 mm. Besa M.G. (in front gunner's compartment). One P.L.M. mtg., with twin Vickers A.A. guns. One Thompson sub-M.G. One 2 in. smoke bombthrower.
Ammunition	6-pdr. ... 75 rounds. Besa M.G. ... 4,950 ,, Vickers A.A. guns ... 2,000 ,, Thompson sub-M.G. ... 320 ,, Bombthrower ... 30 ,,
W/T set	No. 19 set (in rear of turret).
Suspension	10 road wheels independently sprung; 8 shock absorbers (four each side).
Engine	12-cyl. 60° V-type, pressure liquid cooled petrol engine.
Clutch	Dry twin-plate, hydraulically operated, 16 in. diameter (O.D. of driving plates).
Fuel tanks	2—capacity {R.H., 60 galls. L.H., 56 galls.} 1 auxiliary—capacity 30 galls. (fitted).
Oil tanks	2—capacity {R.H., 8½ galls. L.H., 6 galls.
Cooling	1 header tank 2 radiators } capacity 14 galls. (approx.).
Engine starting	Electric starter motor mounted on bevel gearbox.
Range (on acceptance test)	Road, 87-174 miles Cross country, 58-87 miles } without auxiliary tank.
Road speed	40 m.p.h. at 2,550 r.p.m.
Turret traverse	Hydraulic pump driven through bevel gearbox from engine. Traverse motor mounted in turret and geared to turret ring. Auxiliary hand traverse.

Gearbox	Z.5 type. Five forward speeds and reverse.
	 Ratios:—Reverse 22·894 to 1 3rd ... 2·855 to 1
	           1st ... 11·643 to 1 4th ... 1·807 to 1
	           2nd ... 4·593 to 1 5th ... 1·343 to 1

Final drive	Twin-type sprockets each side of hull at rear. Ratio 3·71 to 1.
Steering gears	Epicyclic gears, controlled by hydraulically operated brakes.
Road brakes	Internally expanding type, hydraulically operated by foot pedal.
Length of track on ground	12 ft. 3 in.
Track centres	8 ft. 1¾ in.
Track pressure	13·27 lb. per sq. in.—15½ in. tracks. 14·7 lb. per sq. in.—14 in. tracks.
Vertical obstacle crossed	3 ft. 0 in.
Gap crossed	7 ft. 9 in. (with flat approach).
Height of idler from centre	2 ft. 4¼ in.
Ground clearance	16 in. (approx.).
Fording depth	4 ft. 0 in. (still water with flap closed).
Access doors	Turret roof ... 2 Driver's roof ... 1 Front gunner's roof ... 1
Battery charging	Dynamo, driven through bevel gearbox from engine. Auxiliary charging set in fighting compartment.
Water tanks	12 galls. (fitted in turret).

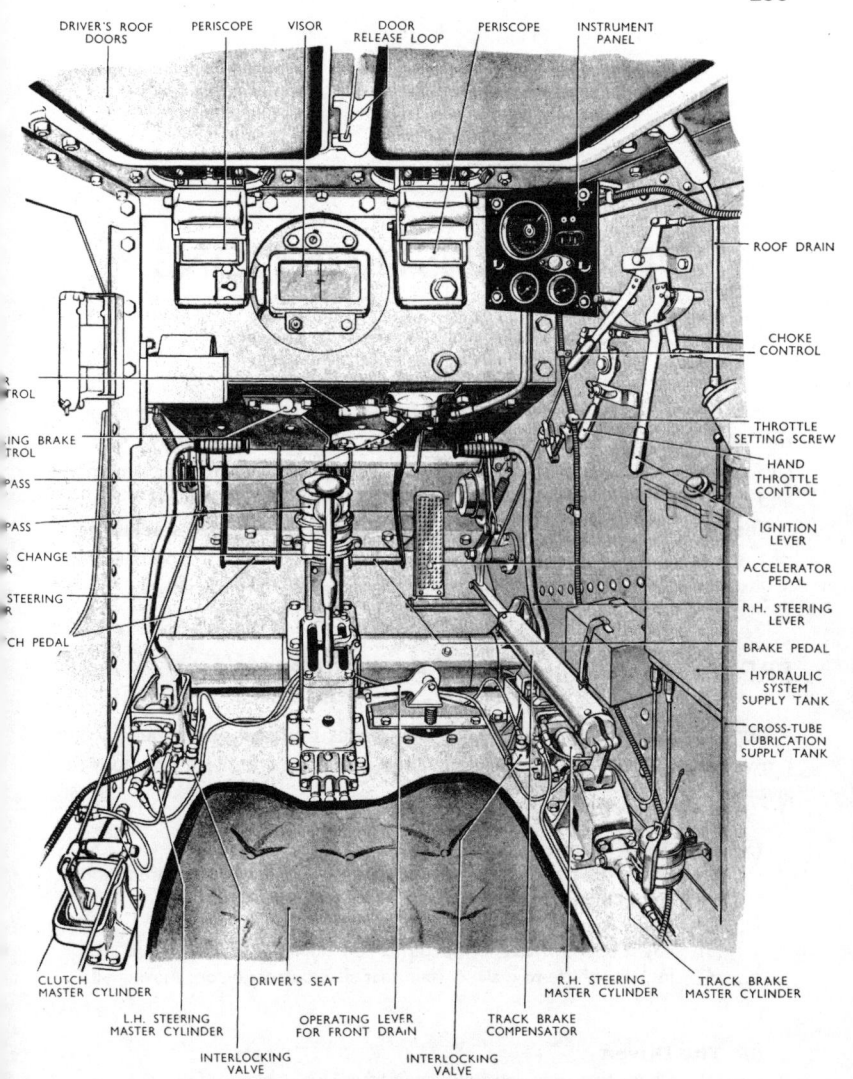

Fig. 3.—Driver's compartment and controls.

DRIVING INSTRUCTIONS

The driver's compartment is situated in the right-hand side of the nose of the vehicle. Access to it is either by the doors in the roof above the seat, or, should these be shut, through the doors in the fighting compartment roof and the hole in the forward compartment bulkhead.

The controls can be brought within easy reach of any driver, by adjustment of the seat position (see Chapter VI A). A general view of the compartment is given in Fig. 3, which should be carefully studied by all new drivers.

PREPARATION OF VEHICLE BY CREW.

When starting up each morning, certain duties must be performed by the crew, and, as these duties are closely connected, the complete procedure is given below:—

(a) **The Tank Commander**.
    (1) Ensure that the vehicle is correctly stowed and that water, rations, ammunition and other equipment are carried as ordered.
    (2) Check that fire extinguishers are full.
    (3) Assist operator to check external communications on W/T set, etc.
    (4) Inspect vehicle to make certain that all is correct before moving off.

(b) **The Driver**.
    (1) Check that main and auxiliary (if fitted) fuel tanks are full.
    (2) Test that coolant and engine oil levels are correct.
    (3) Test that the five flame switches are operating.
    (4) Inspect engine and transmission compartments for oil, fuel and coolant leaks. The floor must be clean and dry. Rectify any leaks and report if normal tightening will not cure.
    (5) Make sure that no paper, rags, waste, etc., is left on track guards on engine compartment roof.
    (6) Instruct front gunner to start engine.
    (7) While engine is running inspect engine again for oil, fuel and coolant leaks. If any leaks are detected, stop engine and proceed again as at (4).

(8) Listen carefully for faulty running. Check both magnetos by instructing gunner to switch off each in turn.
(9) When engine is warm instruct gunner to stop by switching OFF ignition.
(10) Re-check engine oil level and top up if necessary.
(11) Close all engine and transmission covers, make certain that wire handles on front covers are fully down to prevent fouling the turret.
(12) Check that all plates and plugs are in position in hull floor. Instruct wireless operator to open and close rear sump valve, and gunner to work rota-trailer release and auxiliary fuel tank release (if fitted). Check external stowage.
(13) Enter driver's seat, switch OFF exterior lights and close flaps. Start up engine and instruct gunner to traverse turret in both directions by hand and power. Set hand throttle to correct idling speed and switch off engine.
(14) Inspect Triplex block and operation of visor and seat.

(c) **The Wireless Operator**.
(1) Start up auxiliary charging set if necessary.
(2) Check W/T sets as instructed in Wireless Manual.
(3) Test turret lights and ventilating fan for correct operation.
(4) Operate rear sump valve when instructed by driver.
(5) Check operation of periscopes and revolver ports. Also test 2-in. bombthrower by cocking and pressing trigger. Dry clean bore if firing is expected.
(6) Stow waterproof sheets and camouflage nets.
(7) Instruct front gunner to switch ON exterior lights and check that all lights are operating correctly.

(d) **The Front Gunner**.
(1) Switch ON battery cut-off switch and yellow warning light will glow.
(2) Remove safety screws from pins of both $C.O._2$ bottles.
(3) Check level of fluid in hydraulic reservoir, and level of oil in suspension tube lubrication reservoir.
(4) Test operation of sump valve in driver's compartment and free movement of both steering levers, clutch pedal and brake pedal.
(5) Test operation of each steering interlock valve, by holding one lever fully back and checking that the movement on the opposite lever does not exceed 5/8 in.
(6) Apply foot-brake and engage locking ratchet.
(7) Make certain that gear lever is in neutral and steering levers fully forward.
(8) Move carburetter strangler control lever *fully* back.
(9) Switch ON ignition, depress clutch pedal and instruct gunner to operate Ki-gass priming pump.
(10) Press starter button and red warning light will glow. Release as soon as engine fires. If engine does not start, release starter button and try again. *When engine is running do not touch steering levers.*
(11) As soon as engine will run without strangling, return carburetter strangler control lever to forward position. *Never use Ki-gass pump on a rearm engine, but careful use of the carburetter strangler control lever is permissible provided care is taken to avoid excess rich mixture.*
(12) Engage clutch pedal, being prepared to switch OFF immediately should vehicle start to "swing" in neutral.

(13) Warm up engine by setting hand throttle to give medium speed.
(14) Check all gauge readings. Engine oil pressure should rise immediately when the engine starts. Test panel and compass lights, also front gunner's light and fan. Set speedometer trip to zero.
(15) When instructed by driver, switch OFF engine, and when instructed by wireless operator, switch ON all exterior lights and open driver's flap.
(16) Enter front gunner's compartment and carry out first parade on gun, ammunition and stowage.
(17) Check that machine gun is secure on its mounting. Prove gun and then test firing gear and oil frictional surfaces as necessary. See that deflector and chute are secure. Dry clean bore if firing is expected.
(18) Test seat and brow-pad and check that telescope and periscope are clean.
(19) Check that all spares, tools, M.G. oil can and plugs are clean and stowed, also that all stowage and ammunition are correct in this compartment.

(e) **The Gunner**.
(1) Turn on light above powered traverse filter, check oil level in recuperator and pump handle until indicator shows "full".
(2) Check that control handle of variflow pump is in OFF position.
(3) Turn ON fuel tap, and when instructed by front gunner, unscrew the Ki-gass knob and inject three pumpfuls of fuel into induction system. *Screw up knob tightly after use.* The use of this pump is only necessary under very cold conditions.
(4) When instructed by driver, depress each magneto emergency stop switch in turn. Set turret speedometer trip to zero.
(5) When instructed by driver, operate rota-trailer release and also auxiliary fuel tank release (if fitted).
(6) Examine 6-pr. breech mechanism for dirt, rust or other defects. Oil lightly, frictional surfaces if necessary. See that all nuts and bolts and pins on the mountings are tight. Test firing gear. See that deflector and chute are secure, also test elevation and depression. See that S.A. cam is at S.A. Examine level of oil in buffer cylinder and top up if necessary, examining for leakage. Dry clean bore if firing is expected, and replace muzzle cover.
(7) Check that co-axial gun is secure in its mounting, prove the gun, then test the firing gear and oil frictional surfaces if necessary. See that deflector and chute are secure. Dry clean the bore if firing is expected.
(8) Test seat, brow-pad, commander's hatch and mounting lock. Check that telescope and periscope are clean.
(9) Prove first, then test action of A/A machine gun. Prove and test Thompson machine gun by cocking and pressing trigger, controlling bolt action. Oil frictional surfaces if necessary. Check the elevation and traverse of the P.L.M. mounting and finally clamp with handle bars horizontal. Dry clean the bore of these weapons if firing is expected, and slightly oil the chamber of the carbine.
(10) See that spare parts and tools, especially the spare striker case, Wesco and M.G. oil cans, clearing plugs and cleaning rods are clean and correctly stowed.
(11) Test powered traverse system for air, and bleed if necessary. Remove all obstructions and, when instructed by driver, traverse once by hand and in each direction by power. With aid of wireless operator, test operation of depression stop while traversing.
(12) Inspect front of engine compartment and bevel gearbox for leaks.

When parking the vehicle, turn OFF the fuel tap and switch OFF the battery cut-off switch.

DRIVING THE VEHICLE.

Learn to handle the vehicle properly on a good road before venturing into rough and open country.

With the engine running:–

(1) See that the carburetter strangler control lever is in the forward position.
(2) Release the parking brake by pressing fully on the brake pedal and pulling ratchet control knob.
(3) Depress clutch pedal, and when this is right down, engage gear required. If any difficulty is experienced in engaging the gear, ease one of the steering levers just sufficiently to take up the free movement only. The gear can then be engaged easily.

Note – First gear should only be used in confined spaces, on steep inclines or when sharp turns are required. Otherwise always start in second gear.

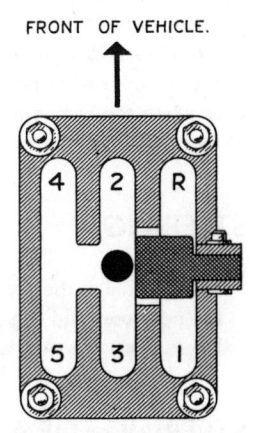

Fig. 4.—Gear change gate.

(4) Slowly accelerate the engine and, at the same time, release clutch pedal. The clutch should be engaged at low engine revs., whenever possible.

Gear-changing:–

Normal double-declutching methods must be used for all gear changing.

When the vehicle is moving, accelerate the engine until the vehicle speed is 5 m.p.h., and engage the next (third) gear. When driving on a level surface the vehicle should be in third gear within 6–10 yards of the starting point. While driving on hard ground make the slightest pause in the gate, but when crossing soft ground move straight through the gate without hesitation. Always use sharp, clean movements when going through the gate.

Do not attempt to change gear when steering. To change from third to fourth gear proceed as above with the vehicle running at 10 m.p.h., and finally change from fourth to top at 15 m.p.h.

Before changing-up while driving across country, and particularly on soft ground, depress the accelerator; then change gear. When stopping on an incline for any period, do not rely solely on the brakes. Stop the engine and engage a gear.

Ignition lever – This lever is not required for use with the Meteor engine and is therefore disconnected.

Accelerator hand lever – This enables the driver to control the engine speed by hand. The lever can be fixed in any given position by tightening the thumb-screw, as when using the power traverse gear with the vehicle stationary.

Slow-running screw – This is provided for increasing the idling speed of the engine which, for normal running, is governed by the carburetter jet settings.

To check oil pressure – The only definite method of checking the minimum engine oil pressure is to drive the vehicle at 9 m.p.h. in second gear when the engine is hot. The oil gauge should then show a pressure of at least 25 lb. per sq. in.

STEERING.

Do not snatch the steering levers, but use a firm steady pull.
Do not steer and use your main brake at the same time.
Do not touch either lever when the engine is idling.

STEERING FORWARD.

The vehicle is steered by altering the relative track speeds, and it follows, therefore, that by increasing the forward speed of the right-hand track the vehicle will turn to the left, and vice versa.

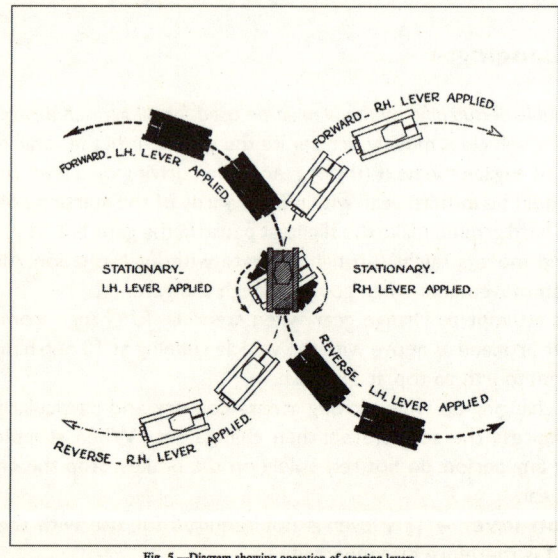

Fig. 5.—Diagram showing operation of steering levers.

Steering control is by means of right and left-hand steering brake levers, one on each side of the driver's seat (*see* Fig. 3).

To turn the vehicle to the *left*, pull back the *left*-hand steering lever. To turn the vehicle to the *right*, pull back the *right*-hand steering lever.

Should the vehicle be moving forward on the higher gears and a sharp turn is requires, the driver must change down to a lower gear.

STEERING IN REVERSE.

Remembering that you are seated *facing forwards*, to cause the rear end of the vehicle to turn to *your* left, pull the *right*-hand lever. For the rear end of the vehicle to move to your *right*, pull the *left*-hand lever. This can more readily be understood on referring to Fig. 5 which shows that no matter in what direction the vehicle is moving, pulling the *right-hand* lever turns its nose to the right and its tail to the left, while pulling the *left-hand* lever turns its nose to the left and its tail to the right.

BRAKING AND STOPPING.

To slow down or stop, apply the foot brake, which actuates the hydraulically operated track brakes fitted between the gearbox output shaft and the final drive. When stopping, declutch and move the gear lever into neutral.

Before getting out of driver's compartment:–

(1) Apply the parking brake (*see* Fig. 3).
(2) Switch OFF engine ignition.
(3) Switch OFF battery cut-off switch.
(4) Turn fuel tap handwheel (in fighting compartment) to OFF position.

STEERING CHARACTERISTICS.

(1) **In gear, clutch disengaged** – Steering effective as long as vehicle is moving, but steering effort will slow down vehicle.
(2) **In neutral, clutch engaged** – Steering effective according to engine r.p.m., whether vehicle stationary or moving. Steering effort will tend to stall engine unless the latter is accelerated.
(3) **In neutral, clutch disengaged** – No steering under any conditions.

Note – If steering is applied while changing gear, it becomes operative immediately the gear is engaged, i.e. before the clutch is engaged (*also see* Chapter IV A).

THE ENGINE

LEADING PARTICULARS

1. **GENERAL.**
   - Type of engine ... Normally aspirated, pressure-liquid-cooled V-engine.
   - Number of cylinders ... 12.
   - Arrangement of cylinders ... Two lines of six cylinders forming a 60° V.
   - Bore ... 5·4 in. (137·16 mm.).
   - Stroke ... 6·0 in. (152·4 mm.).
   - Capacity ... 1,649 cu. in. (27 litres).
   - Compression ratio ... 6 to 1.
   - R.A.C. rating ... 140 h.p.
   - B.H.P. ... 570 to 600 at 2,550 r.p.m.
   - Torque ... 1,450 lb./ft. at 1,500 r.p.m.
   - Cylinder numbering ... From fan drive end—1A, 2A, 3A, 4A, 5A, 6A, 1B, 2B, 3B, 4B, 5B, 6B.
   - Firing order ... 1A, 2B, 5A, 4B, 3A, 1B, 6A, 5B, 2A, 3B, 4A, 6B.
   - Nominal governed speed ... 2,550 r.p.m.

2. **FUEL.**
   - Type ... 67 octane upwards (Pool).
   - Consumption ... Road—·75 to 1·5 miles per gallon.
     Cross country—·5 to ·75 miles per gallon.

3. **OIL.**
   - Type ... 10 H.D. (M.160).
   - Consumption ... 40 miles per gallon (approx.).
   - Pressures—
     - Main ... Minimum—35 lb. per sq. in. at 2,550 r.p.m.
     - Auxiliary ... Minimum—3 lb. per sq. in.

4. **IGNITION.**
   - Number and type of magnetos ... Two F.S.T./12/R, G/2.
   - Number per cylinder and type of sparking plug ... Two RC76/2.
   - Magneto timing—
     - Fully retarded ... Inlet 5° before T.D.C.
       Exhaust 10° before T.D.C.
   - Contact breaker gap ... 0·012 in.
   - Sparking plug gap ... 0·012 in.

5. **CARBURATION.**
   - Carburetters ... Two 56 – DC.

6. **VALVES.**
   - Valve timing ... Inlet opens 7° before T.D.C.
     Inlet closes 45° after B.D.C.
     Exhaust opens 45° before B.D.C.
     Exhaust closes 7° after T.D.C.
   - Valve clearance ... 0·030 in. for timing.
     0·020 in. for running.

7. **COOLANT.**
   - Type ... 100% pure water or pure water plus ethylene-glycol (see "Frost Precautions").
   - Maximum outlet temperature ... 110° C. – (230° F.).

8. **STARTING SYSTEM.**
   - Type ... Electric turning gear.

263

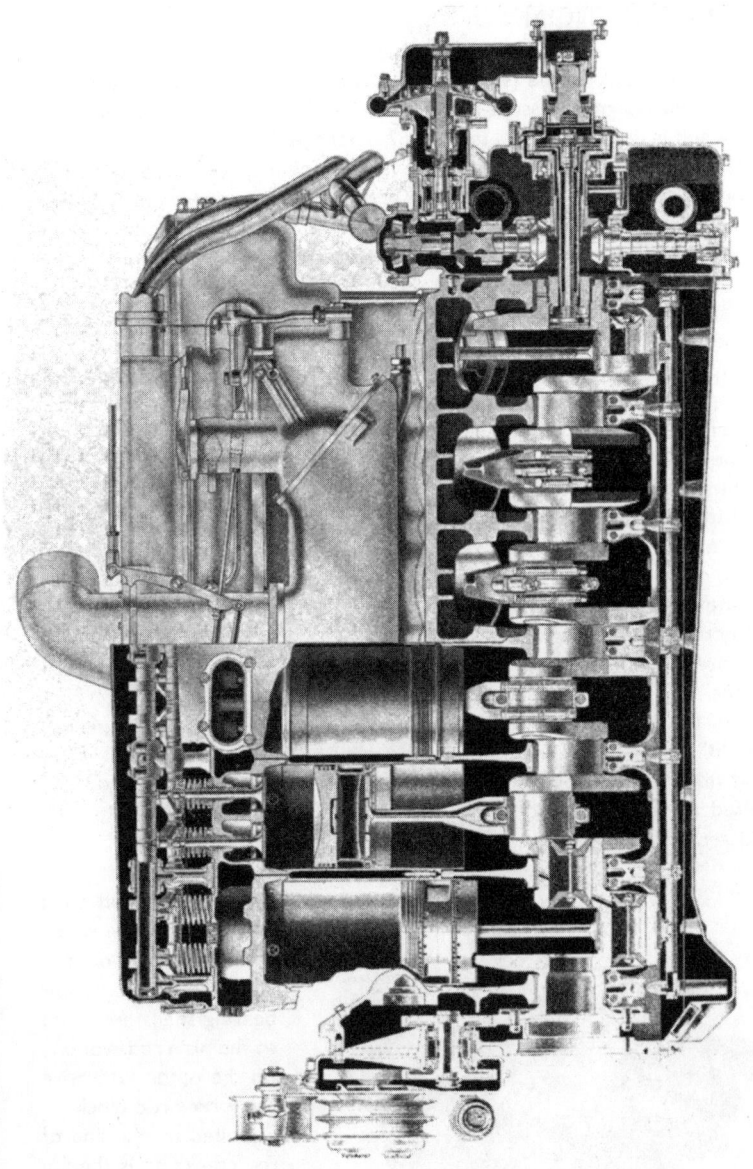

Fig. 9.—Section through engine.

INTRODUCTION

The Meteor tank-engine (see Figs. 9, 10 and 11) is of the 12-cylinder type, having two integral pressure-liquid-cooled banks of six cylinders forming a 60° V, and is designed to operate on 67 octane fuel.

The two monobloc non-detachable-head cylinder blocks are mounted on inclined upper faces of the crankcase, and are designated "A" and "B" banks respectively, the "A" bank being on the left-hand side of the engine when viewed from the fan drive end. The "A" (left-hand) side camshaft cover bears the cylinder firing order plate. The separate steel liners are of the wet type and are provided with shoulders at each end which abut against the cylinder block and crankcase respectively. Each cylinder has *four* valves – two inlet and two exhaust valves and two sparking plugs. The valves of each cylinder block are operated from single centrally-disposed overhead camshafts through a system of individual tappet fingers.

The balanced six-throw crankshaft is supported within the crankcase in seven lead-bronze lined main bearings. The connecting rods are H-section steel forgings and are of the plain type on the "A" (left-hand) side, and of the forked type on the "B" (right-hand) side. A divided steel block is bolted to the forked rod and retains a flanged thin

Fig. 10.—Meteor I—left-hand or "A" side.

lead-bronze lined steel shell in its bore, which works directly on the crankpin. Similar split bearing shells are fitted to the plain rod, working on the outer surface of the forked rod block.

Bolted to one end of the crankcase is the fan drive unit and mounted at the other end of it is the wheelcase which houses the gear wheel assemblies transmitting

Fig. 11.—Meteor I—right-hand or "B" side.

the drive to the camshafts and wheelcase accessories. The gears are driven from the crankshaft through a torsionally flexible shaft which acts as a spring drive. Twisting of this shaft is positively limited, and torsional oscillations are damped by a special friction drive. The shaft also serves to smooth out irregularities in angular velocity and torque in the drive between the crankshaft and auxiliary components. The wheelcase provides mountings for the magnetos, coolant pump, fuel pumps and oil pumps. An electric starter, mounted vertically in the Fighting Compartment adjacent to the engine bulkhead, is connected to the outer end of the spring drive shaft by a circular flanged coupling.

Carburation is provided by two twin-choke up-draught type carburetters, located below the induction manifolds between the cylinder blocks, fuel being supplied to them from the pumps mounted one on each side of the wheelcase.

The lubrication system is of the dry sump type, and one pressure pump and two scavenge pumps are employed. These three pumps are mounted on the wheelcase, the two scavenge pumps on the "B" (right-hand) side and the pressure pump on the "A" (left-hand) side of this unit. One scavenge pump drains the crankcase and the other the wheelcase. A proportion of the main pressure oil is transformed by a reducing relief valve to low-pressure, for the purpose of lubricating the fan drive unit, camshafts and wheelcase drives.

The final drive is transmitted by means of a shaft, the forward end of which is splined to the crankshaft. Midway along this shaft is a circular flange to which is bolted the clutch. An electrical governor incorporated in each magneto prevents the maximum engine speed from being exceeded. The magnetos also carry an automatic timing device for advancing the spark as the engine speed is increased.

MAINTENANCE.

Maintenance on the Meteor engine has been cut to a minimum under the workshop overhaul period, and is confined to certain items of the ignition, fuel, oil and cooling systems which are dealt with under their appropriate headings. Such things as tightness of nuts, pipe unions and hose connections must be checked and engine control linkages oiled EVERY 250 MILES.

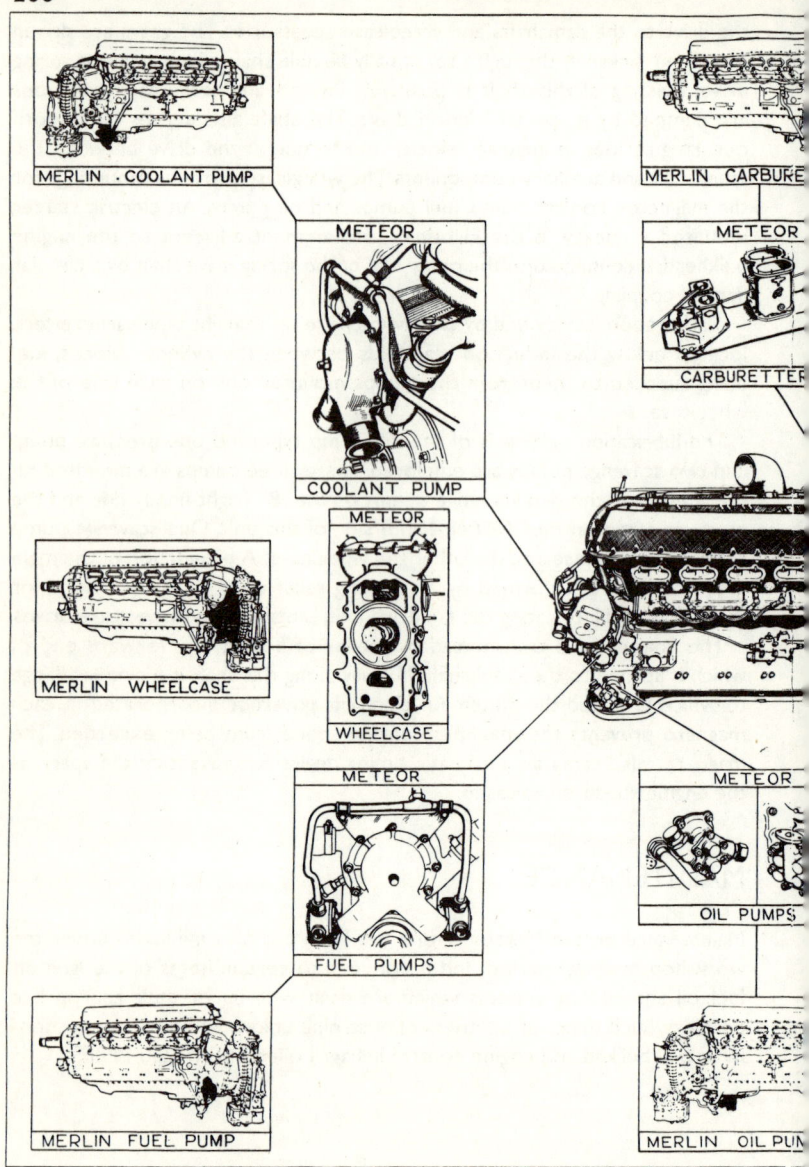

266

MERLIN COOLANT PUMP

MERLIN CARBURE[TTOR]

METEOR

METEOR

COOLANT PUMP

CARBURETTER

MERLIN WHEELCASE

METEOR

WHEELCASE

METEOR

FUEL PUMPS

METEOR

OIL PUMPS

MERLIN FUEL PUMP

MERLIN OIL PUM[P]

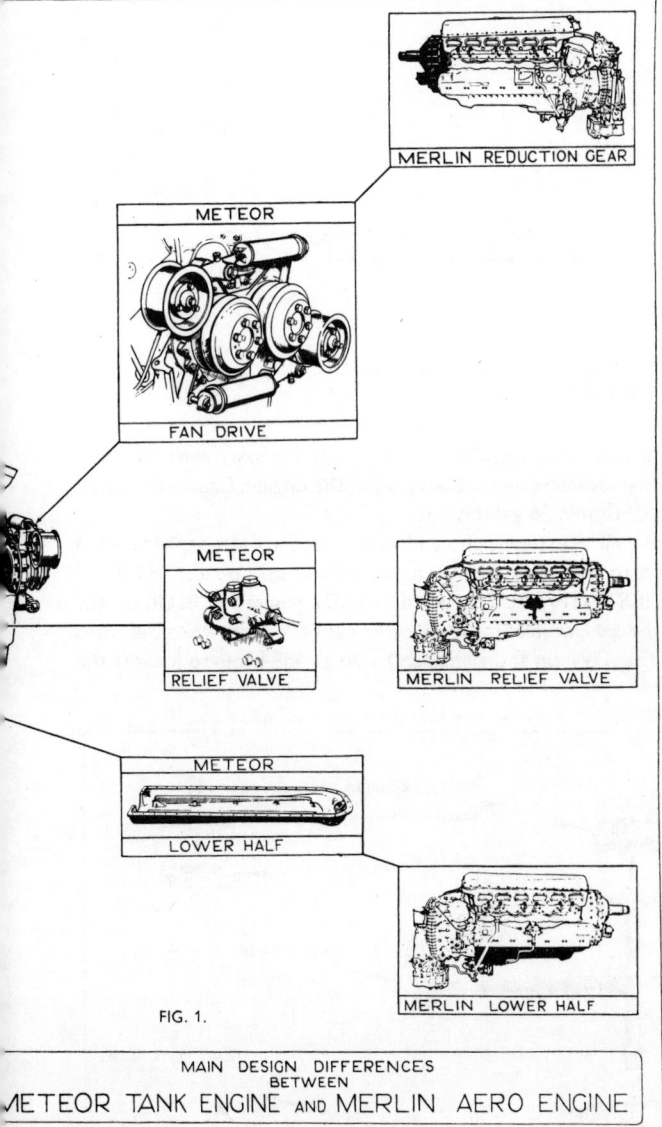

267

MERLIN REDUCTION GEAR

METEOR

FAN DRIVE

METEOR

RELIEF VALVE

MERLIN RELIEF VALVE

METEOR

LOWER HALF

MERLIN LOWER HALF

FIG. 1.

MAIN DESIGN DIFFERENCES
BETWEEN
METEOR TANK ENGINE AND MERLIN AERO ENGINE

FUEL SYSTEM

MAIN FUEL TANKS.

Fuel is carried in two tanks installed in the engine compartment (see Figs. 124 and 125), being located one on each side of the engine. Capacities – R.H. tank, 60 gallons; L.H. tank, 56 gallons.

The filler caps are pressure-tight and are vented to atmosphere via a pipe which is led to the space between the vehicle side plates. DO NOT FILL THESE TANKS WITH OIL. The word PETROL painted in BLUE on the inclined face of the side air inlet louvre locates each filler cap, while the word OIL painted in YELLOW on the right-hand side air inlet louvre locates the oil filler cap.

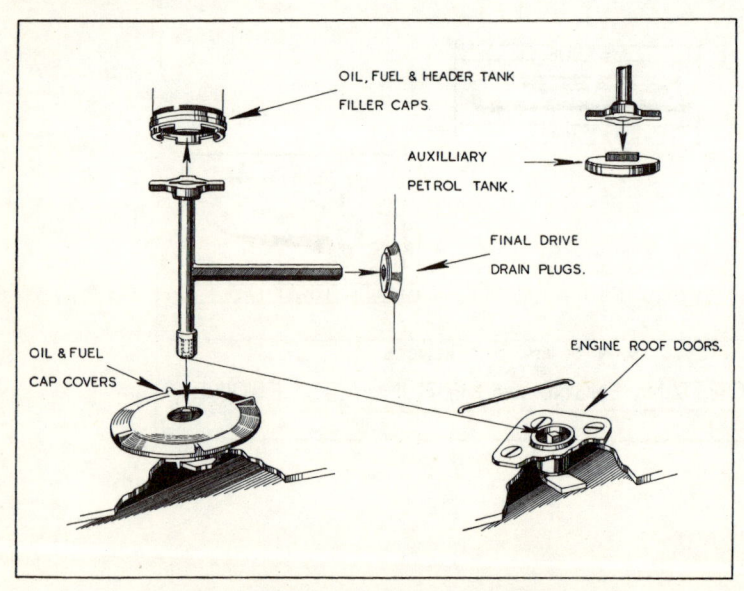

Fig. 13.—Composite tool for doors, covers and filler caps.

Filling – Each main fuel tank is filled in the usual manner by removing the access cover and unscrewing the filler cap (see Fig. 13). Insert the hose into the filler neck and carry on filling until the fuel level in each tank (no balance pipe is fitted) is up to the bottom of the filler neck. Replace filler caps and bleed the system.

Draining – Fig. 67 shows the position of the fuel tank drains and outlet pipe covers. Only remove the drain plugs when it is necessary either to completely drain the tanks of fuel or to drain away any water that may have collected in them. In the latter case the plugs need not be fully unscrewed and should be screwed up again as soon as fuel starts draining from the tanks.

The outlet pipe covers are detached only when the tanks are lifted, or when for some other reason the outlet pipes require attention.

AUXILIARY FUEL TANK (if fitted).

An auxiliary fuel tank, containing 30 gallons, may be fitted to the rear of the vehicle (see Fig. 126), in which case the fuel in it should be used when moving to the scene of operations so that, if necessary, the tank may be jettisoned. To jettison the tank, pull the handle in the right-hand rear corner of the fighting compartment. This action releases the two straps retaining the tank in position, so that it falls away under its own weight and at the same time disconnects its supply pipe at the rubber junction.

To replace the tank, push in the handle, lift the tank into position by the handles provided and close the two straps. A pin fixed to the end of each strap fits into a catch, which is then snapped into position. Refit the rubber tube to the supply pipe.

FOUR-WAY TAP.

All three fuel tanks are connected to a four-way tap mounted on the left-hand side of the intermediate bulkhead, and is operated by a handwheel in the fighting compartment (see Fig. 128). This wheel is inscribed with the word PETROL at the bottom, and four other markings which, when turned so that they coincide with a pointer, are as follows:–

(1) OFF– indicates fuel supply shut off.
(2) R.H – indicates right-hand fuel tank in use.
(3) L.H – indicates left-hand fuel tank in use.
(4) AUX – indicates auxiliary fuel tank in use.

Movement of the handwheel to the appropriate position is all that is necessary, as this makes the required connection. Fuel gauges are mounted

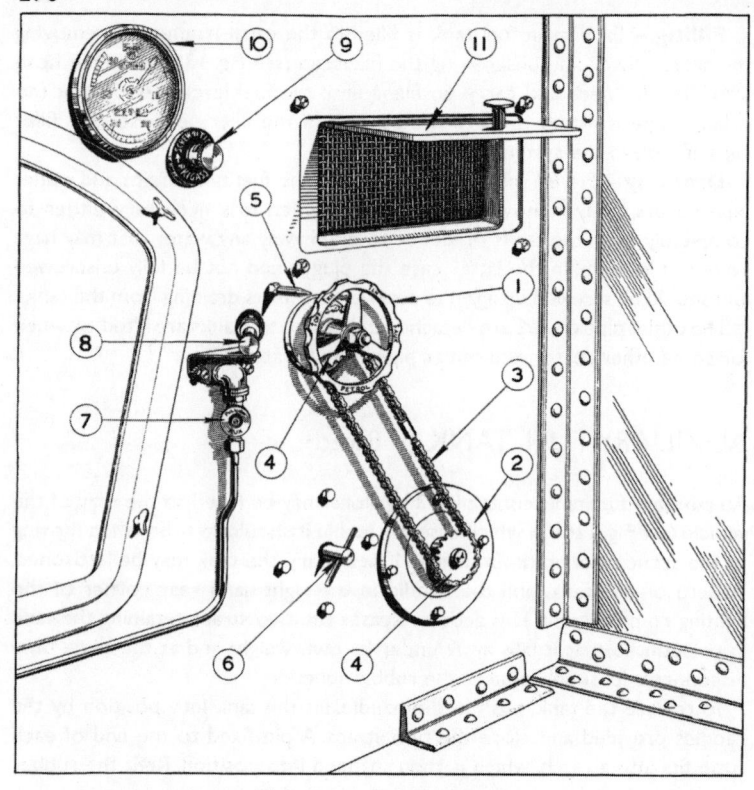

Fig. 128.—Fuel system controls.

1. Hand-wheel.
2. Four-way tap unit.
3. Bicycle-type chain.
4. Sprockets for (3).
5. Pointer for (1).
6. Filter tommy bar.
7. Fuel control for auxiliary charging motor.
8. Air-bleed control for filter.
9. Ki-gass priming pump.
10. Speedometer.
11. Intake for air cleaner.

on the driver's instrument panel for each of the main fuel tanks, but none is provided for the auxiliary tank.

OIL SYSTEM

The oil for lubricating the engine is stored in two interconnected tanks located one on each side of the engine (see Figs. 134 and 136).

A filler cap is provided in the right-hand tank which is reached after removing the access cover in the engine roof (see Fig. 13). The word OIL is painted in YELLOW on the inclined face of the right-hand air-intake louvre adjacent to this filler cap. DO NOT POUR PETROL INTO THIS TANK.

The two tanks are connected by means of a balance pipe so only one filler cap is provided, but a dipstick is fitted to the left-hand tank (see Part 5, Chapter I B).

FILLING.

(1) Pour oil into the filler tube in the right-hand tank until the FULL mark on the dipstick in the left-hand tank has been reached. Pour quickly so as to ensure that the right-hand tank is full by the time the FULL level in the left-hand tank is reached.
(2) Replace filler cap.
(3) Replace dipstick.

DRAINING.

Each tank has a separate drain plug, and these can be reached by removing the oil tank drain covers in the floor plates (see Fig. 67). The plugs can then be unscrewed (see Fig. 136), but when doing so, be certain to have a container underneath to prevent wastage. The total capacity is 14½ gallons, so use a receptacle which is large enough. Most of the system can be drained through one tank plug, but removal of the other plug is necessary to clear the residue contained below the level of the balance pipe. Drain the system when the engine oil is *warm* and with the vehicle on *level* ground. The plugs can also be used for draining any water in the system. To do this, unscrew each plug in turn until oil starts to flow and then tighten the plugs.

CHANGING OIL.

(1) It is preferable to change over to the new H.D. oils while the engine is in a new condition, or as soon after an overhaul as possible. Engines which are nearing overhaul should, if possible, continue to run on the existing mineral grades of oils (see Lubrication Chart).

(2) When changing over to the H.D. oils, the engine must be run until the normal working temperature is attained and the existing oil drained off while hot.

(3) The engine oil tanks should then be filled (as above) with the appropriate grade of H.D. oil and the engine run at approximately half throttle for half an hour. The oil should then be drained off again while hot, and all easily removed filter elements, gauzes, pipe lines and relief valves must then be removed and cleaned.

(4) The engine oil tanks must then be refilled with the H.D. oil and a label affixed to the driver's instrument panel near to the oil pressure gauge, giving in large letters the grade of H.D. oil used, the date and mileage of the change-over; this information must also be entered in the A.B.413.

(5) Normal running should then be continued for not less than 300 or more than 500 miles, during which time special notice should be taken for evidence of low oil pressure or overheating of the engine. If a drop in oil pressure or overheating is experienced, the engine must immediately be stopped, the oil drained off, and the lubrication system cleaned as in (3) above, and replaceable filter elements renewed.

(6) If the operation of the engine has been normal during the 300–500 mile run, the oil must still be drained off while hot and the lubrication system cleaned as in (3) above and replaceable filter elements renewed. The label on the instrument panel and the A.B.413 must be endorsed to the effect that the second draining of the oil has been carried out.

(7) Normal oil changes and operation of the vehicle can then be resumed.

COOLING SYSTEM

The cooling of the radiators depends entirely upon the circulation of air by two fans. These fans are built into the front of the two radiators which are installed one on each side of the rear end of the engine compartment (see Part 6, Chapter I B, for description and illustrations). One header tank serves as a reservoir for the system, the total capacity being approximately 14 gallons.

The coolant is circulated by means of an engine-driven centrifugal pump and is pumped through the cylinder block and induction pipe coolant-jackets, and through a thermostat with by-pass, into the header tank. From the header tank it travels downwards through the radiators where it is cooled by the forced draught of air from the fans. From the bottom of the radiators it is drawn up into the circulating pump and so to the engine again.

To work efficiently the cooling system must be kept clean and full of coolant. The level must therefore be checked DAILY. If the system has to be topped up daily, check for leaks.

Filling the system – The system is filled through the header tank, by lifting the access door on top of the engine compartment (see Fig. 5A) and unscrewing the filler cap. *Do not remove the filler- cap while the engine is running, and wait until the temperature falls below 100° C. (212° F.).*

When filling, the level will rise slowly until about half-way up the filter, when it will rise rapidly. At this point the system is full. Run the engine for about half a minute, then top-up.

Do not, at any time, remove the filter except for cleaning. If the coolant takes a long time to run through the filter, the gauze needs attention. Clean same by removing the complete filter cage and swilling it in petrol.

After filling – Screw home the filler cap sufficiently tight to make a pressure-tight joint. Do not overstrain in tightening, but it is not sufficient to have the filler cap just hand-tight.

Draining the system – The system is drained by means of an evacuation pump mounted at the rear of the header tank (see Fig. 145), thus simplifying the preservation of coolant when operating far from adequate water supplies or when using anti-freeze mixture (see "Frost Precautions"). The vehicle should be run on to level ground before operating the pump.

The pump is connected by piping to a three-way tap, and this in turn is connected to the sumps of each radiator. A length of flexible hose is supplied for attachment to the pump outlet, and the loose end of this pipe should be placed in a suitable clean container (see Fig. 16).

The tap is marked R.H., L.H. and OFF. Turn the tap to R.H. to drain the right-hand radiator and to L.H. to drain the left-hand radiator, and work the ram up and down. Great care must be taken to see that the system is pumped completely dry, with the tap in both positions. Collect the coolant in the container and keep covered until it is poured back into the system.

RADIATORS.

Remove inspection covers in fan cowls EVERY 1,000 MILES and remove foreign matter which has lodged against the matrices. Provided the radiators are free from oil, they are unlikely to become choked, but if in an oily condition they will rapidly become blocked up, in which case the only effective remedy is to have them removed and degreased.

TRACKS AND TENSIONERS

TRACKS.

The tracks are driven by sprocket wheels at the rear of the vehicle, the latter having double teeth which engage the manganesed-steel links of the track.

A new track consists of 125 links, but it is necessary to add one spare link, making a total of 126, for the initial "running in" period. This is only a matter of a few miles' running, after which the spare link can be removed and each track will then number 125 links until such time as it becomes necessary to remove another link, due to normal wear, in order to keep the track correctly tensioned. This may occur several times, at varying intervals, during the life of a track.

Each link (see Fig. 22) has a tread or "spud" to engage the ground and a positioning lug or "horn" is formed on the inner face to centre the track on the driving sprockets, the road wheels and the track tensioner wheels.

The links are hinged together by hardened-steel pins. A head is formed at the inner end of these pins, while their outer ends are reduced in diameter to take a thick washer over which the end of the pin is riveted. This thick washer must be fitted on the pin, small chamfer first, so that the pin end can be riveted over into the large chamfer (see Fig. 22). A loose washer is interposed between the thick washer and the link. This construction is clearly shown in Fig. 22.

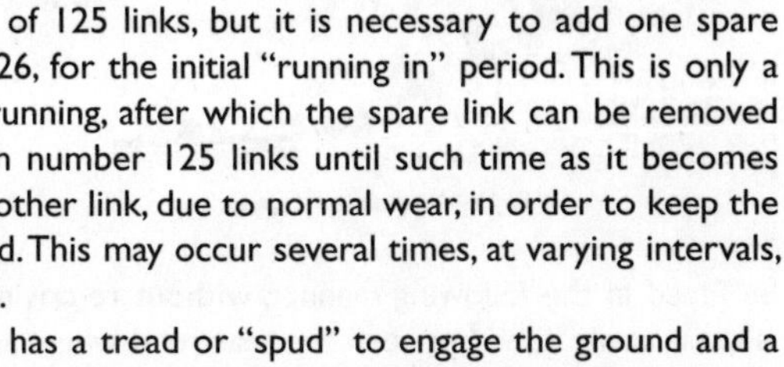

The head of the pin is always placed

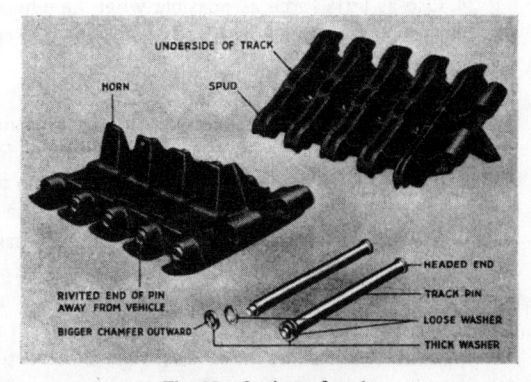

Fig. 22.—Sections of track.

towards the vehicle so that the riveting can be done on *the outside*, away from the vehicle. It is most important that the track be fitted in the manner shown in Fig. 23, that is, with the *spuds trailing as they pass round the sprockets*.

MAINTENANCE.

Do not lubricate the tracks at all, but keep them as clean as possible. Examine tracks DAILY for loose pins.

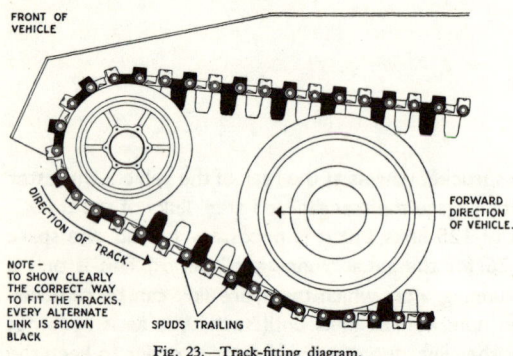

Fig. 23.—Track-fitting diagram.

Replacing a Track Pin – When the track has been working under arduous conditions, it is occasionally found that the thick washers (*see* Fig. 22) become loose or even drop off, so that the pin will commence to work out. A new pin can be fitted in the following manner, without removing the track, and the job is most easily done between the rear road wheel and the driving sprocket.

The pin which is to be replaced will generally be part-way out of its hole. Insert an old discarded pin into the hole and knock the pin right out. This leaves the track held in position by the old pin which is being used as a tool. Insert a new pin and proceed to knock out the "tool" pin in the opposite direction, that is, away from the vehicle. When this operation is complete, fit the loose washer and then the thick washer and rivet the pin over as in Fig. 24. Use as little force as possible when hammering the new pin home, or otherwise the reduced diameter end will be burred and it will be difficult to fit the two washers.

To remove a Track – Under certain circumstances it may be necessary to remove a track so that new links can be fitted. To do this, first run the vehicle on to firm level ground, if practicable. Slacken the track in the manner described under "Adjustment of Track Tension", and with the gear lever in neutral and foot-brake off, rotate the sprocket by means of a crowbar so that the bottom of the track is slack between the sprocket and the rear road wheel.

Knock out a link pin by placing a drift against it at the riveted end, and hitting the drift sharply with a heavy hammer. Then rotate the sprocket in the opposite direction so as to disengage that portion of the track which is around it.

With the crew positioned between each road wheel at the side of the vehicle, drag the top run of the track over the idler wheel and lay it on the ground in front of the vehicle.

Replace the damaged links, driving out the link pins as described above. Insert new pins by riveting over as shown in Fig. 24. Take care to support the pin head as shown in B, Fig. 24, when riveting the pin end.

To replace a Track – With the track repaired and the crew arranged alongside the vehicle as before, drag the top run of the track over the idler wheel and along the road wheels. Attach a wire rope to the top run rear link and weave the rope in and out of the sprocket teeth. Again using the crowbar, rotate the sprocket so that the track is pulled over the sprocket teeth and meets the other end of the track.

Support both link ends and replace the pin. Fit the two washers and rivet over the end of the pin as shown in Fig. 24. Finally, adjust the track tension.

A broken track can be repaired and replaced in this way provided it is not completely shed.

To fit a new Track – Proceed in the manner described under "To remove a Track", until the old track is on the ground with the top run pointing forward.

Place the new track, which must be exactly the same length, in line with the rear of the old track, i.e. pointing rearward. Both tracks should then form a single line, the new one being a continuation of the old one.

The vehicle must now be driven in reverse gear so that the new track is under the road wheels with the larger section lying forward.

As only one track is fitted, it will be necessary to lock the free sprocket before the vehicle will move. To do this, pull the steering lever on the *same* side as the fitted track. Then use the clutch to drive the vehicle backwards.

Drive the vehicle very slowly until the road wheels are over the new track. Then complete the operation as described under "To replace a Track".

A   Fig. 24.—Riveting a track pin.   B

TRACK TENSIONER

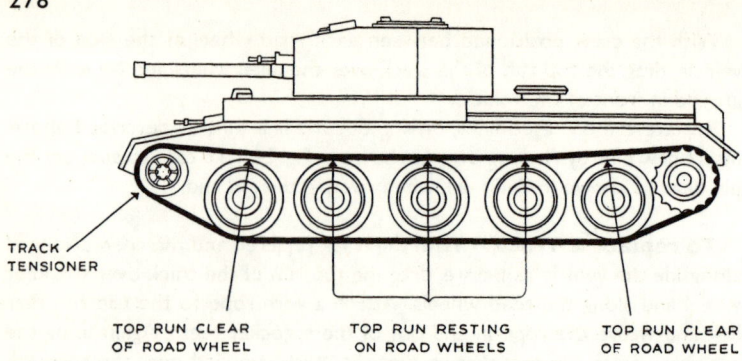

TOP RUN CLEAR OF ROAD WHEEL

TOP RUN RESTING ON ROAD WHEELS

TOP RUN CLEAR OF ROAD WHEEL

Fig. 25.—Track correctly adjusted.

Note – Where possible, it is quicker to procure a tow from another vehicle, as it will be appreciated that the method described above is, of necessity, rather slow.

Should a track be shed, it must be repaired and refitted as described above, i.e., the track should be placed on the ground behind the vehicle and in direct line with the road wheels.

TENSIONERS.

The track must always be properly tensioned as shown in Fig. 25, and to check this the vehicle should be run on to flat ground. The top run of the track should be clear of the front and rear road wheels, but should rest lightly on the three centre wheels.

Adjustment of each track is provided for by tensioners fitted at the front of the vehicle. Each tensioner (*see* Fig. 183) consists of a rubber-tyred wheel running on ball and roller bearings mounted on an eccentric axle. This axle is pivoted in a fixed housing bolted between the inner and outer hull plates of the vehicle.

The tensioners (*see* Fig. 183) are fitted with a ratchet wheel (1) and a pawl arm (3) by which the ratchet wheel is locked in position. The pawl arm is itself secured by the lock-nut (5). Two slots are provided for insertion of the end of the adjusting lever which is carried on the right-hand side of the engine compartment roof.

Adjustment of Track Tension – To disengage the ratchet wheel, first unscrew the lock-nut (5) so as to release the spring-loaded pawl arm. Then insert the end of the adjusting lever into one of the two slots and lift upwards to tighten the track. There is no danger of the tensioner slipping backwards and so slackening the track as the pawl arm (3) is spring-loaded into engagement with the ratchet wheel (1) and will "click" round over it.

Check to see that the correct tension has been achieved as shown in Fig. 25, and then tighten up the lock-nut (5).

MAINTENANCE.

The tensioners are grease lubricated and should be greased EVERY 250 MILES through the nipples incorporated in the idler wheel hubs. The eccentric axles must also be greased EVERY 250 MILES, the two grease nipples being located on the right-hand side of the front gunner's compartment, the grease being forced along piping to each eccentric axle at points (8), (Fig. 183).

STEERING BRAKES AND TRACK BRAKES

Fig. 27.—Steering and track brakes.

There are two sets of brakes:– **THE STEERING BRAKES** which are attached to the gearbox, and **THE TRACK BRAKES** which are anchored to the hull sides (see Fig. 27). Both are generally similar in design and are hydraulically operated and totally enclosed as protection against oil and dirt. The reserve tank for the hydraulic fluid is on the right-hand side of the driver's compartment.

STEERING BRAKES.

The steering brakes are applied by hand levers in the driver's compartment. The *left-hand* lever controls the *right-hand* brake and the *right-hand* lever controls the *left-hand* brake. These levers should be applied smoothly and

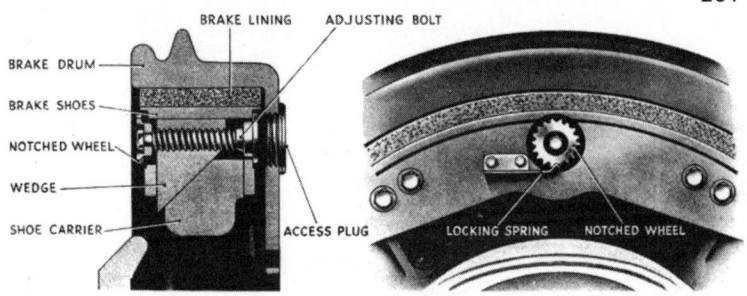

Fig. 28.—Wedge adjustment of brakes.

gently. Harsh application gives erratic steering, particularly when travelling at high speeds on the road.

When driving on the lower gears, the sharper the turn the greater the effort required to operate the steering levers.

Do not apply the steering controls when the vehicle is stationary and the engine running, even when the gear lever is in neutral, unless a pivot turn is needed. If a pivot turn to the left is required pull the left-hand lever. The left-hand track then moves backwards while the right-hand track moves forwards, so that the vehicle turns to the left about its own centre. Pulling the right-hand lever has the opposite effect and the vehicle pivots to the right. Therefore *take care* when touching either steering lever while the engine is running and the vehicle is stationary.

Skid turns are only possible when reverse gear is engaged, when pulling the right-hand lever locks the left-hand track and vice versa (see "Driving Instructions").

TRACK BRAKES.

The track brakes are used for slowing down the vehicle or bringing it to rest and are operated by the foot pedal in the driver's compartment.

When "parking" the vehicle press the foot pedal hard down, at the same time *pushing* the ratchet control knob until the ratchet teeth engage the pedal. The pedal will then stay down and the brake ON. To release the brake press hard on the foot pedal, and if the ratchet does not spring out of engagement *pull* the knob to disengage it.

THE HULL

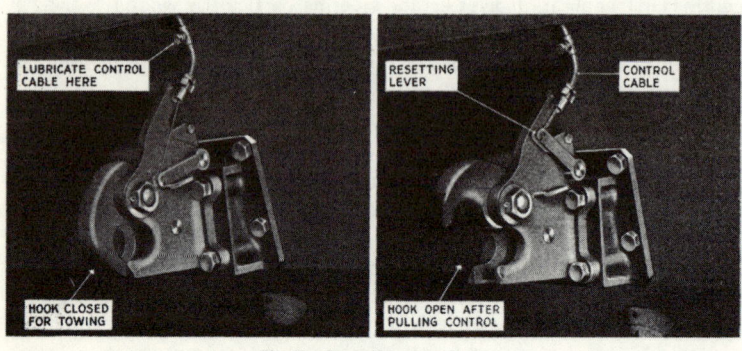

Fig. 61.—Rotatrailer towing hook.

The hull is divided into four compartments:—

(1) **The Driver's Compartment** – This is in the right-hand part of the nose of the vehicle.
(2) **The Front Gunner's Compartment** – In the left-hand part of the nose of the vehicle.
(3) **The Fighting Compartment** – This is amidships, below the turret, and houses the rotating platform which bears the commander, gunner and loader-operator.

Towing eyes are provided at both the front and rear ends of the vehicle, while lifting eyes are mounted on top of the nose and also at the top of the inner sideplates at the rear of the vehicle.

A special hook for towing a rotatrailer is fitted at the rear of the vehicle (see Fig. 61). This is operated by cable from a handle in the top left-hand rear corner of the fighting compartment (see Fig. 62). By pulling this handle the hook is opened and the trailer released. The hook can only be re-set from outside the vehicle.

DRIVER'S COMPARTMENT.

The driver's compartment houses the instrument panel, compass, steering-brake levers, track-brake pedal and all driving controls. For instructions in the use of the instruments and driving, see "Driving Instructions".

Access to the compartment is either by the door in the roof above the driver's seat, or should these be shut, through the doors in the turret roof and then through the hole in the bulkhead at the back of the driver's compartment.

To open driver's doors from inside the vehicle (or from outside through visor) pull the chain which runs along the leading edge of the right-hand door. This withdraws the spring-loaded fasteners and enables the driver to swing the doors upwards.

NOTE – *Never close all doors when leaving the vehicle as they cannot be opened from outside if properly shut.* Always leave the driver's visor open, or the commander's cupola doors in such a position (i.e. with one door overlapping the other), that the driver's doors or commander's cupola can be opened from outside the vehicle.

Engine Controls – These are three in number, are grouped on the right-hand side of the driver's compartment and are easily accessible. They comprise an accelerator hand lever, and a carburetter strangler lever. The third lever, which is disconnected on this vehicle, is for use with the "Liberty" engine for ignition timing.

(1) *Accelerator* – This can be operated either by hand or foot. When it is necessary for the driver to operate both clutch and brake pedals at the same time, as when starting on a hill, the hand lever is used. It can be fixed in position by tightening the thumb-screw on the lever.

(2) *Strangler lever* – This operates the valves in the strangler valve body of the induction system and is used in conjunction with the Ki-gass pump for cold starting conditions. It is spring-loaded to the "forward" or "fully open" position. (*See* "Driving Instructions".)

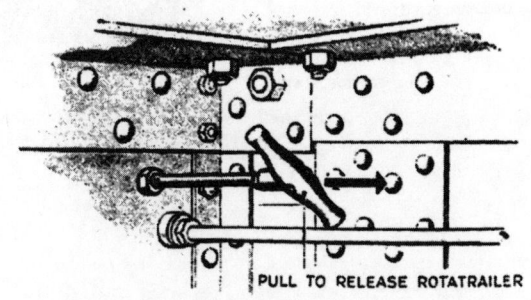

Fig. 62.—Rotatrailer hook control.

FRONT BULKHEAD.

OUTER SIDE PLATE.

FIGHTING COMPART

DRIVER'S COMPARTMENT

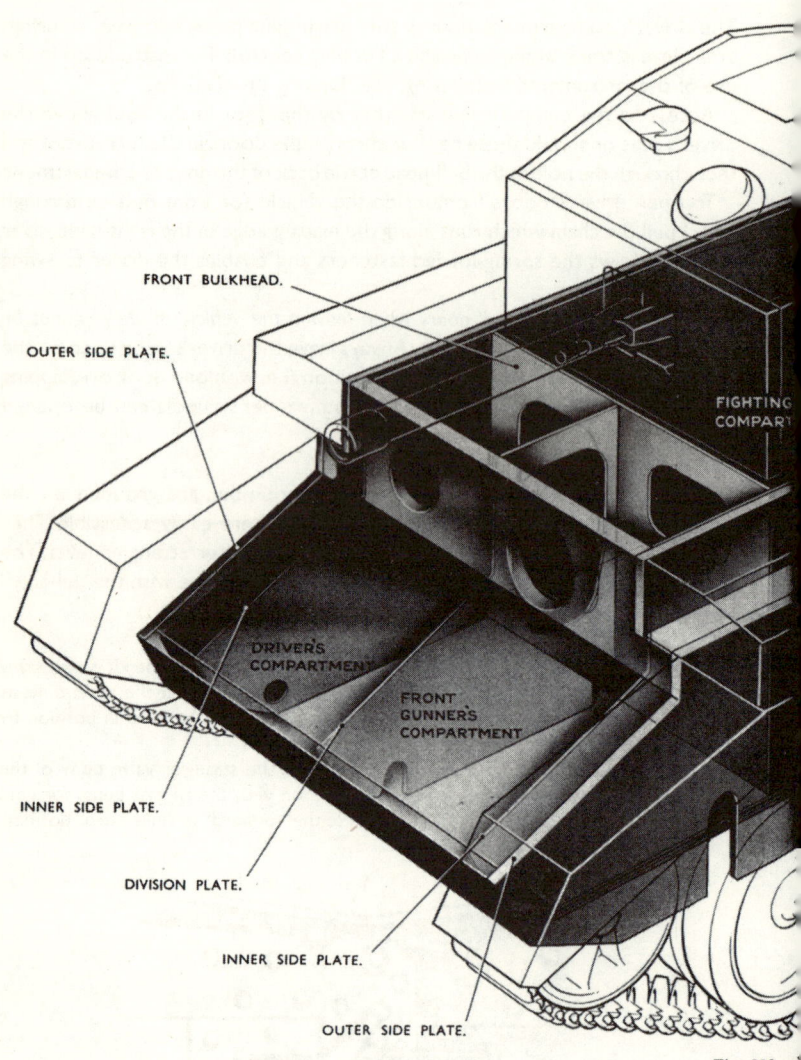

FRONT GUNNER'S COMPARTMENT

INNER SIDE PLATE.

DIVISION PLATE.

INNER SIDE PLATE.

OUTER SIDE PLATE.

Fig. 228.—

Fig. 228.

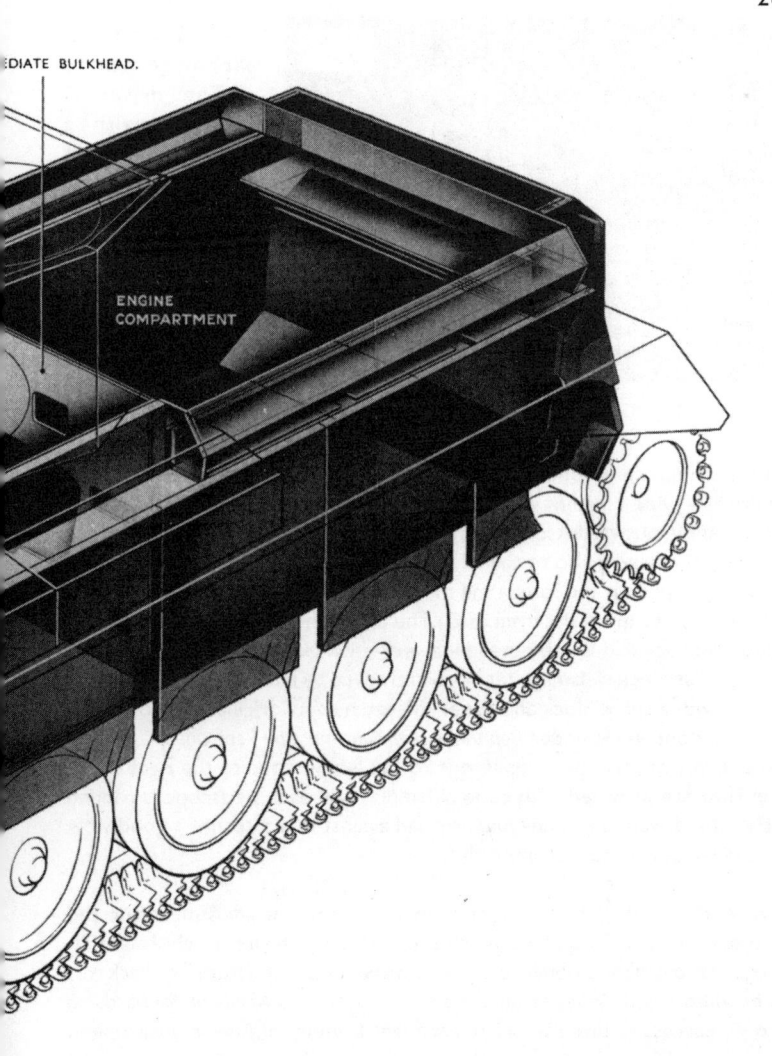

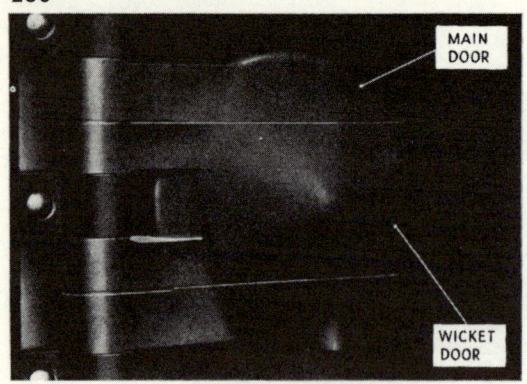

Fig. 63.—Driver's visor closed.

Vision Arrangements – The driver is provided with a visor in the front and two periscopes in the roof. This visor is armoured and incorporates a small wicket door and a main door. The wicket door can be opened and the main door left closed; or the complete unit can be opened when the vehicle is not under fire. This gives the driver a wide range of vision. To open the complete unit, pull the spring-loaded handle outwards (see Fig. 3), move the main catch to the left in its slot and swing the handle also to the left. To open the wicket door only, press the operating plunger, at the same time moving the handle as before. In this case be careful not to move the main catch. The driver then has a smaller aperture to look through and is protected by a very thick glass block.

If this glass block is damaged, it can be removed by lifting the securing catch. Replace with a spare block and lower the catch to its original position. If this has to be done while under fire *the wicket door must be kept closed*.

The driver has two periscopes, one to the left and one to the right of the visor. They are provided with control handles and each periscope is padded so that the driver can steady his forehead against it and secure a good view. For further details, see Chapter VII A.

Driver's Seat – This is mounted on slides providing adjustment fore and aft, to suit the driver's comfort. By releasing a catch at the front right-hand side, a range of movement is obtained to suit drivers of varying stature. The back rest can be lifted slightly in its sockets and then hinged backwards or forwards so as to give access to the driving compartment from the fighting compartment.

Front Sludge Drain – On the right, in front of the driver, is the front sludge drain in the hull floor (see Fig. 3). This is operated from inside the vehicle only. To open the drain, lift the cam lever and

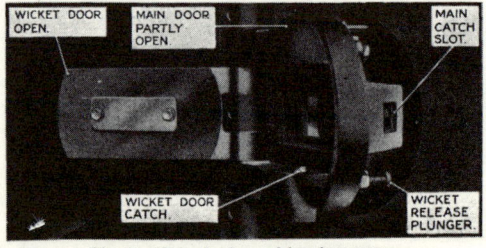

Fig. 64.—Driver's visor, wicket door open.

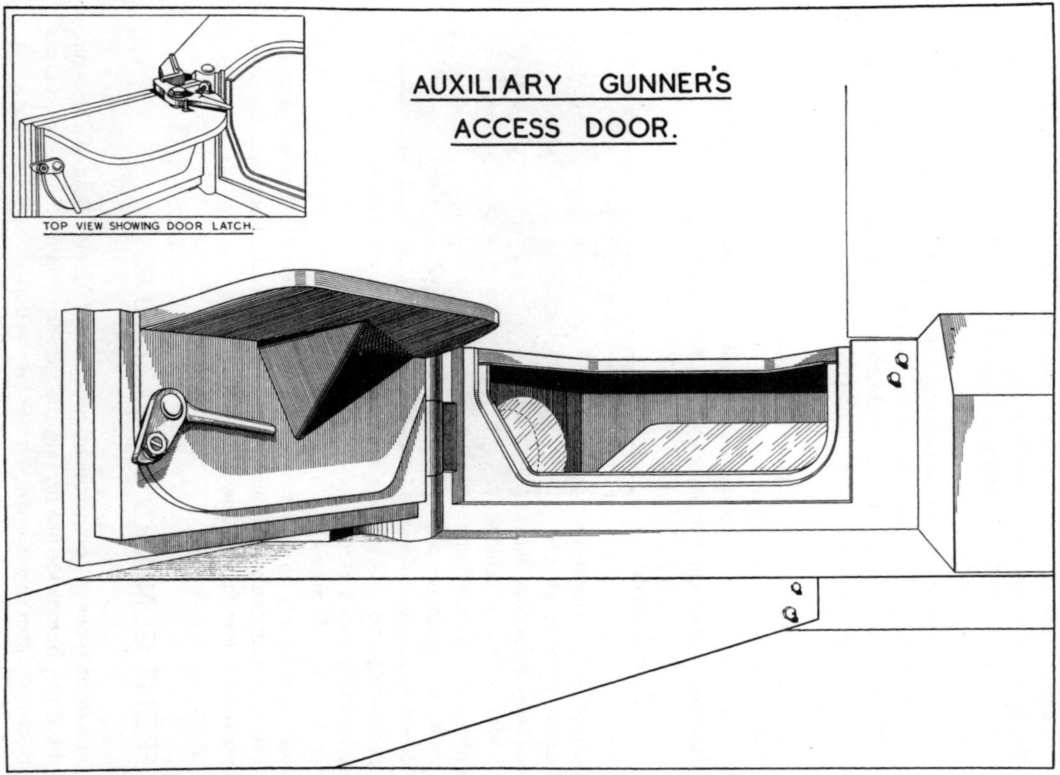

Fig. 65.—Front gunner's access door.

swing it hard over to the right. When the lever is lifted over to the left again, the drain is closed by a spring. Keep the mechanism clean and oil the cam lever pivot when required.

Lubrication – Keep the driver's roof doors in good working order by WEEKLY inspection and oiling when required.

The reserve tank for the hydraulic brakes is on the right-hand wall and this should be checked WEEKLY and topped up as required.

The reserve tank for the axle-arm lubrication system, also on the right-hand wall of the driver's compartment, should be topped up EVERY 250 MILES.

There is one nipple at the front of the change-speed box and three nipples on the change-speed rods on the floor. These should receive attention EVERY 250 MILES.

Each periscope has two nipples on the circular slide. Lubricate these EVERY 250 MILES, but do not overdo this.

The seat slides and the various controls in the compartment should be inspected and oiled if necessary EVERY 250 MILES.

Maintenance – The glass block in the driver's visor and also the periscope lenses must be cleaned by wiping with a wet rag and polishing with a soft cloth. Where mud has hardened on, wet it thoroughly before removing to avoid scratching the surface of the lenses. To prevent fogging, apply a small amount of anti-dim compound No. 2 to the glass. If a lens is broken the pieces must be thoroughly brushed out with the brush provided before fitting a new one.

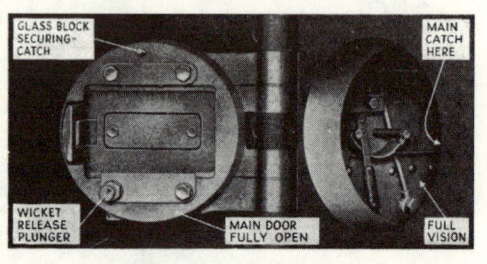

Fig. 66.—Driver's visor, fully open.

FRONT GUNNER'S COMPARTMENT.

The front gunner's compartment contains a 7.92 mm. Besa machine-gun, the $C.O._2$ bottles for fire fighting, the axle-arm lubrication pumps, and an electrically-operated extraction fan for ventilation.

Access to this compartment is through the front gunner's side door (see Fig. 65) or through the doors in the turret roof in the same manner as outlined for the driver's compartment, and then through the hole provided in the division plate.

Vision Arrangements – No periscope is provided for the front gunner, but a sighting telescope is fitted for use with the Besa machine-gun.

Pumps for Axle-Arm Lubrication System – The cross-tubes carry the bearings for the axle-arm pivot shafts, and these bearings are lubricated by pressure pumps mounted on the floor, to the right of the front gunner's compartment. The pumps draw oil from the reserve tank on the right-hand wall of the driver's compartment, and deliver it under pressure to all the cross-tube bearings. They are automatically-operated each time the driver depresses the clutch pedal (see Fig. 227). As the amount of oil to be delivered to each bearing has been established by test, *do not make any alteration*.

For Seating, Lubrication and Maintenance, see remarks under "Driver's Compartment".

FIGHTING COMPARTMENT.

The fighting compartment extends the full interior width of the vehicle. The front bulkhead separates it from the driver's and front gunner's compartment, while the rear bulkhead forms a division between this and the engine compartment. Extending downwards into this compartment from the turret and rotating with it is the turntable which carries the commander, gunner, and loader-operator.

On the right-hand wall at the front of this compartment is the electrical control board, and below it the main switch-box. Below this, anchored to the floor, is the auxiliary charging set. The batteries are cradled on the same wall at the rear of the compartment.

In the rear corner by the left-hand wall is the recuperator for the power-traverse, and nearby the remote control fuel-tap is mounted on the rear bulkhead (see Fig. 127) and also the Ki-gass fuel pump and commander's speedometer.

In the centre of the rear bulkhead are the magneto emergency stop switches (see Fig. 49), while at either side are the air-cleaner doors. Near the floor is the bevel gearbox with the starter-motor on top, the dynamo on the right-hand side and the power traverse variable-flow pump on the left-hand side.

The operating handle for opening the rotatrailer hook is in the top rear corner above the power traverse recuperator, while the handle for jettisoning the auxiliary fuel tank (if fitted) is in the opposite top rear corner above the batteries. The base-junction for the power traverse is mounted in the centre of this compartment and a metal cover is fitted to the turntable platform to surround and protect it.

Drains – The centre drain is slightly to the right-hand rear of the fighting compartment and must be opened when draining the power traverse system.

It is reached by removing the detachable board in the turntable platform. To open the drain, push the handle downwards; to keep it open, give the handle a half-turn. The drain is closed by a spring (see Fig. 60).

The rear sludge drain is at the rear of the vehicle floor, but is operated by a lever in the fighting compartment – on the left-hand side of the rear bulkhead (see Figs. 49 and 60). To open the drain, lift up the lever. The drain is self-closing under pressure of a spring.

THE TURRET

The turret, in which the 6-Pr. gun and 7.92 mm. Besa machine gun are co-axially mounted, is heavily armoured and can be completely rotated so that the guns can be brought to bear upon a target without having to turn the vehicle. Doors are provided in the turret roof, those on the right-hand side giving access to the loader-operator's position, while the doors in the commander's cupola on the left admit to the commander's position and gunner's seat.

Four periscopes are fitted in the roof, one for the gunner and one for the loader-operator, and two in the cupola for the commander. In addition, an electrically-operated ventilating fan is mounted in the forward part of the roof and two wireless aerials are carried on the rear part of the roof. In the right-hand forward corner of the roof a bracket is spigot-mounted. This bracket is in the form of an oblique tube and is flanged at its lower end to receive a 2-in. smoke bombthrower which is bolted to it.

Revolver ports are provided in the turret rear sides and a sighting telescope for the gunner is mounted co-axially with the guns.

TURRET ROTATION AND LOCK.

Both manual and power rotation is provided, the traversing gear being dealt with in detail in Chapter VIII B.

A travelling lock is fitted on the turret ring so that the turret can be locked solid in any desired position, and consists of a spring-loaded plunger operated by a cam-lever (see Fig. 69). The plunger has teeth to engage the

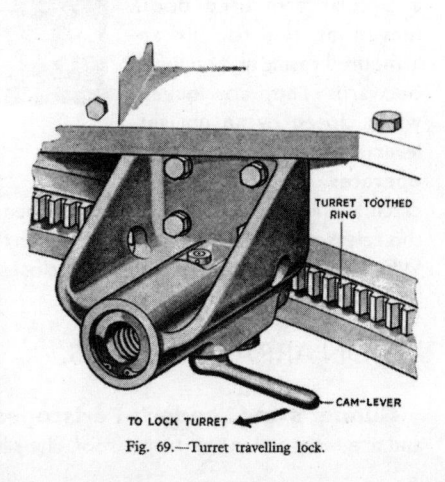

Fig. 69.—Turret travelling lock.

turret ring teeth, thus ensuring a positive lock. When the turret has been traversed to the desired position, the lock must be engaged by moving the handle hard over to the left.

The lock must be *completely disengaged* by releasing the handle hard over to the stop, *before* operating the power traverse – otherwise damage will occur.

COMMANDER'S CUPOLA.

This is the rotating armoured cover in the turret roof, which rotates on three rollers carried on ball-bearings bolted to the cover itself. It is manually operated by two pairs of traversing handles, and can be locked in four definite positions by means of a spring-loaded bolt. A locking-handle, operating on one of the doors, permits both doors to be locked from inside. Two spring-loaded catches are provided outside the cupola to hold the doors open.

LOADER'S DOORS.

These provide the alternative means of entering the turret. To open them from inside the vehicle, pull the chain which runs along the leading edge. This withdraws the spring-loaded fasteners and enables the doors to be swung upwards.

REVOLVER PORTS.

Two ports are provided, one in each rear side of the turret (see Fig. 70). Each consists of a circular armoured door, hinged at the top in an armoured casing and opening outwards. They are locked when closed by an upright lever inside the turret which operates a spring-loaded catch at the bottom of each port. To open the port, pull the lever inwards; this releases the catch and a further pull on the lever swings the port upwards. When the lever is released, the port closes and locks itself.

Fig. 70.—Revolver port.

VISION ARRANGEMENTS.

Gunner's and Loader's Periscopes – These periscopes are identical and are mounted in the turret roof. The same type of periscope is fitted in the hull roof for the driver. Each is mounted in a flange bolted to the roof plate and protected by an armoured ring with a sheet-metal top cover. The mounting allows each periscope to be rotated and tilted, and a handle is provided for controlling these movements (see Fig. 71).

A slide is provided for back-laying, so that vision can be obtained from either rear or front of the periscope without having to rotate it. This slide has three positions and is controlled by a spring-loaded ball and cup. In the highest position, as shown in Fig. 72, a normal view is obtained, in the middle position both main and back-laying lenses are covered, while in the lowest position the back-laying lens is brought into use. To replace damaged lenses, pull up the small release catch at the front and break the periscope as shown in Fig. 72. The lenses can then be removed and replaced.

Commander's Periscopes – These periscopes are mounted in the commander's cupola, and are internally similar to the gunner's and loader's periscopes, with the same lenses and back-laying arrangement. The top lenses, however, are protected by swivelling armoured covers, operated by levers inside the cupola. The periscopes are mounted for tilting only, any rotation required being obtainable by turning the cupola itself. Damaged lenses can be replaced in exactly the same manner as in the case of the other periscopes.

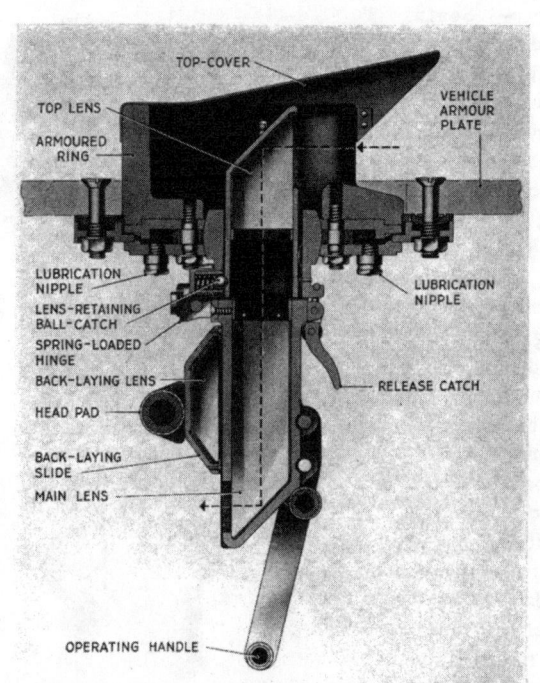

Fig. 71.—Section through driver's, gunner's and loader's periscope.

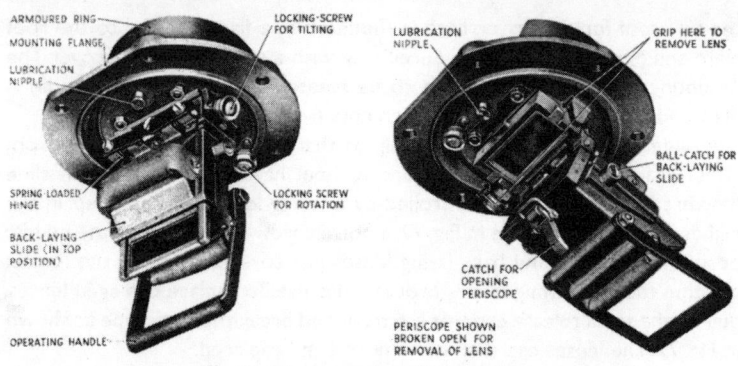

Fig. 72.—Driver's, gunner's and loader's periscope.

Gunner's Telescope – A gun sighting telescope is mounted coaxially with the 6-pdr. and the Besa guns, and is described in Chapter IX B.

TURRET TURNTABLE.

Seating – Seats are provided for the commander, gunner and loader-operator. The commander can only be seated when the cupola doors are open, but when these are closed it is necessary for him to stand on the platform. By lifting his seat slightly and pushing it over to the rear, it will serve as a back-rest.

The gunner's seat is mounted on a pedestal which is bolted to the platform. It is quickly adjustable by means of a foot-pedal. The seat pillar is spring-loaded and when the pedal is depressed a locking-plunger is disengaged

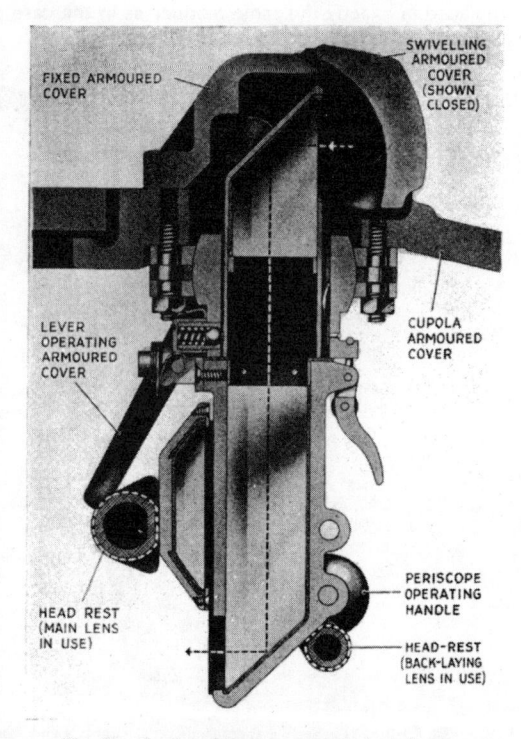

Fig. 73.—Section through commander's periscope.

and the seat can be raised or lowered as desired. When the pedal is released, the locking-plunger reengages and the seat is locked in position.

The loader-operator's seat can be hinged over out of the way when not in use.

WIRELESS AND INTER-COMMUNICATION EQUIPMENT.

The wireless and inter-communication equipment is fully dealt with in a separate book, "Wireless Sets, No. 19, Working Instructions", obtainable from C.O.O., Donnington.

Q.F., 6-PR. 7-CWT. GUN

The Q.F., 6-pr. 7 cwt. guns used in armoured fighting vehicles are adapted from anti-tank weapons to suit A.F.V. gun-mountings. The main difference is in the breech rings, which, in the ground pattern guns, are drilled and tapped for securing to the slippers, and in the A.F.V. patterns, are formed with a lug to connect to the recoil systems. The breech mechanism components are interchangeable.

PARTICULARS.

	Mk. III.	Mk. V.
Weight with breech mechanism	6 cwt. 3 qr. 5 lb.	6 cwt. 1 qr. 27 ½ lb.
Weight without breech mechanism	6 cwt. 0 qr. 1 lb.	5 cwt. 2 qr. 23 ½ lb.
Length – total	100.95 in.	116.95 in.
Length of barrel	96.2 in.	112.2 in.
Calibre	2.244 in.	2.244 in.
Rifling	24 plain section grooves.	24 plain section grooves.
Twist of rifling	1 turn in 30 calibres.	1 turn in 30 calibres.
Probable life of rifling in full rounds, Cordite W., W.M. or W.M.T	800.	
Probable life of rifling in full rounds, N.H.033	1,600.	
Muzzle velocity, full charge, Cordite W., W.M.,		
W.M.T. or N.H.033	2,675 f.p.s. (approx.).	

Muzzle velocity, H.V	2,800 f.p.s. (approx.).	
Firing mechanism	Percussion.	Percussion.
Number of rounds a gun may fire before requiring examination (equivalent full charge)	180.	

OPERATION.

To load – Set the semi-automatic gear to S.A. fire. Open the breech, using the breech mechanism lever.

Return the lever to the "breech closed" position. Insert the round, pushing it into the chamber with a sharp movement. The breech will automatically close.

To fire – Squeeze the trigger of the remote firing control gear.

To unload – Open the breech slowly, using the breech mechanism lever. Extract the round carefully, preventing it from falling from the breech opening with the free hand. Replace the round in the ammunition rack. Ease the firing mechanism by pressing the firing lever on the gun with the right hand and controlling the cocking handle with the left hand.

Misfire –

ACTION BY GUNNER.	ACTION BY LOADER.
Tap the loader twice, shout "Misfire – re-cock".	Re-cock by pulling the cocking handle to the rear with two fingers of the left hand or by using the cocking lanyard. Tap the gunner once.
Squeeze the trigger.	Watch movement of the striker mechanism. If the gun fails to fire, tap the gunner twice.
WAIT FOR ONE MINUTE.	
	Open the breech slowly and examine the cap. If the cap has been struck, throw the round outside the vehicle. If not struck, replace the round in the ammunition rack and change the firing pin.

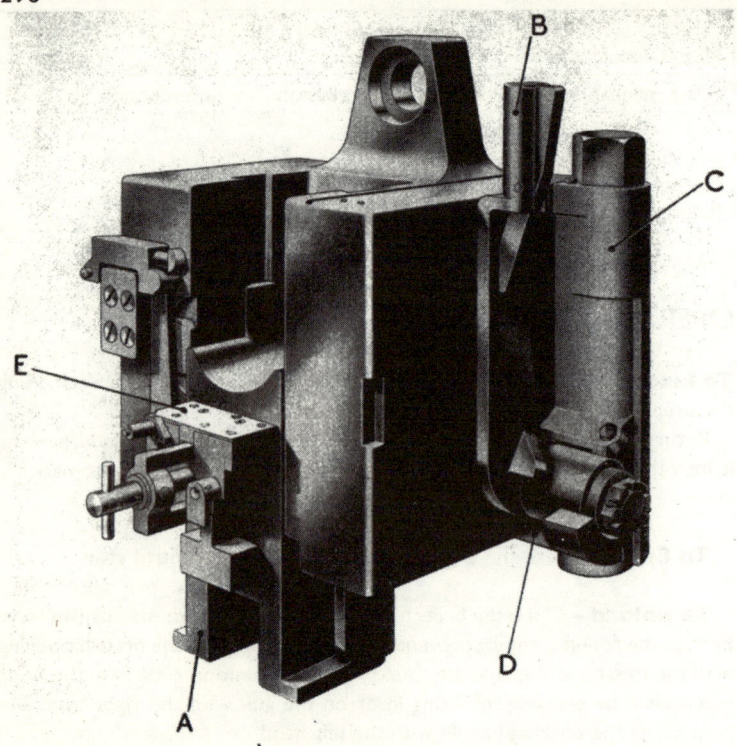

Fig. 76.—Breech mechanism—open.

A. Breech block.
B. Breech mechanism lever.
C. Spring case.
D. Rack pinion.
E. Striker case.

TO DISMANTLE THE BREECH MECHANISM
(see Fig. 76).

- (a) Remove the deflector and loader's shield.
- (b) With the striker cocked and the safety catch at "SAFE", remove the striker case (E) by withdrawing the retaining catch and turning the case through an angle of 60 degrees clockwise.
- (c) Place safety catch at "FIRE" and press sear inwards to release the striker.
- (d) See that the gun is level and remove gun lug nut.
- (e) Remove the actuating shaft keep pin and slotted nut, hold in position the breech block and two extractor levers, and withdraw the shaft towards the left until the breech mechanism lever (B) and rack pinion (D) can be removed. Support breech block (A) during completion of the withdrawal movement of the actuating shaft, then lower breech block a little, push up and remove extractor levers and then withdraw breech block. Remove cocking link actuating pin, crank and cocking links.

TO ASSEMBLE THE BREECH MECHANISM.

(a) Assemble the breech block, with crank, striker cocking link and actuating pin. Insert the breech block a short distance into the breech ring, holding in position the two extractor levers, and place the block in the closed position, making sure that the striker cocking link is flush with the rear face of the breech block. Align the crank and extractor levers and insert the actuating shaft from the left. Place in position the breech mechanism lever and rack pinion, push home the actuating shaft and secure it with the nut and split pin.

(b) Cock the striker and insert the striker case into the breech block. Place the safety catch to "SAFE" and turn the case through an angle of 60 degrees, counter-clockwise.

(c) Open the breech and adjust the compression of the closing spring by releasing the check screw and turning the spring case cap, until the breech can be closed easily without undue slamming, with a dummy round in the breech. When the correct compression is obtained, turn the check screw to lock the spring case cap.

(d) Replace the deflector and the loader's shield, if these parts have been removed.

TO DISMANTLE THE FIRING MECHANISM.

(a) Remove the striker case from breech block and release the striker (**see** "To Dismantle the Breech Mechanism" (b)).

(b) Rotate the safety catch to the "FIRE" position. Grasp the cocking handle in one hand and the case in the other and press the toe of the trigger-sear to ease the main-spring.

(c) Remove the keep pin from the cocking handle and unscrew. Withdraw the cocking sleeve from the rear, and the spindle with the main-spring, from the front of the case. Remove the staple from the head of the spindle and withdraw the firing pin.

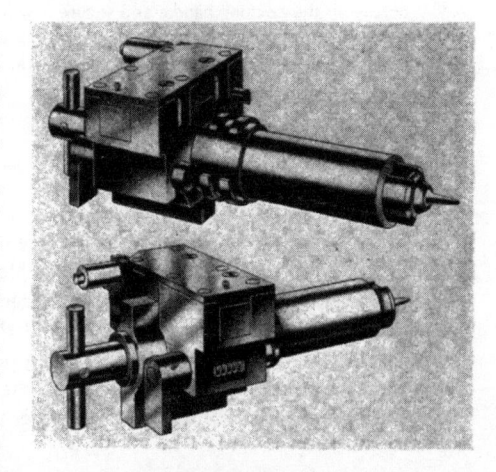

Fig. 77.—Striker case.

(d) Remove the safety catch retaining pin from the top of the case. Withdraw the safety catch to the rear. Take out the split pin from the spindle portion of the catch and remove the plunger and spring.

(e) Remove the split pin securing the trigger-sear spring seat and withdraw the sear and spring. Withdraw the split pins from the roller axis pins and remove the axis pins and rollers.

(f) Remove the split pin and head of the striker case retaining catch plunger and withdraw the plunger and spring from the front of the case.

TO ASSEMBLE THE FIRING MECHANISM.

(a) Insert the retaining catch plunger with the spring, in the striker case, fit on the head of the retaining catch and secure it with a split pin.
(b) Place in position the trigger-sear rollers and axis pins and secure with split pins.
(c) Insert the trigger-sear with seat and spring. The seat and spring fit in the right side of the case and are retained by a split pin.
(d) Fit the spring and plunger in the safety catch. Press the sear inwards and insert the safety catch into the striker case from the rear and set it to the "FIRE" position. Place in position the safety catch retaining pin from the top of the case.
(e) Insert the arm of the cocking sleeve in its recess to engage the slot in the trigger-sear. Press on the left end of the trigger-sear while pushing home the cocking sleeve.
(f) Fit the firing pin in the head of the striker spindle and secure it with the retaining staple. Place the main-spring in position over the spindle, insert in the front of the case and press home to engage the key in the cocking sleeve. Screw up the cocking handle and secure it with a split pin.
(g) Cock the striker and turn the safety catch to the "SAFE" position.
(h) Insert the striker case into the breech block. Set safety catch to "FIRE" and release the striker.

PROTRUSION OF THE STRIKER.

Remove the breech block from the gun (see "To Dismantle the Breech Mechanism" (e)). With the striker case assembled and the firing pin in the fired position, apply the striker protrusion gauge No. 16, to the face of the breech block. The minimum should foul and the maximum should clear the firing pin. If not correct, change the firing pin by dismantling the firing mechanism and removing the firing pin with the drift No. 18 inserted into the hole behind the head of the striker spindle.

EXTRACTION OF JAMMED CARTRIDGE.

If a cartridge jams in the chamber and cannot be extracted by leverage on the B.M.L., it must be removed by using the Key, Removing, Jammed Cartridges, No. 9. Remove the primer with the key, reverse the key and screw it into the primer recess. The cartridge can then be withdrawn by screwing the extracting nut on the key. Misfired rounds must be set aside for examination.

GUN, MACHINE, BESA, 7-92 MM., MARKS I, II, II*, III, III*

The 7.92 mm. Besa Machine Gun used in A.F.V.'s is an air-cooled gas-operated weapon with buffered action, ammunition being supplied by a belt holding 225 rimless cartridges. The mark is stamped on the left-hand side of the gun body.

The Mk. I gun is a converted ground pattern gun for use in A.F.V.'s.

The Mk. II gun is made as an A.F.V. weapon ⎫
The Mk. III gun has a fixed high rate of fire ⎬ See Fig. 79.
The Mk. III* has fixed low rate of fire ⎭

The 7.92 Besa M.G. is intended for mounting in A.F.V.'s, has no ground mounting or sights, and aiming is carried out by means of a sighting telescope housed in the gun mounting.

The barrel cannot be changed unless the gun is removed from the A.F.V. mounting.

The gun can be fired dismounted provided the ejection opening is clear of the ground.

Reference numbers on the illustrations referred to in the following description of the Besa M.G. are for identification purposes, common to the Instruction Book for Armourers D.D. (E) 263.

PARTICULARS.			
Approximate weight complete	48 lb. (varies according to mark).		
Approximate weight of barrel	15 lb. (varies according to mark).		
Overhaul length	3 ft. 7½ in.		
Length of barrel with flash eliminator ...	2 ft. 5 in.		
Rates of fire (rounds per minute):–	High.	Low.	

Mk. I	750/850	450/550 ⎫	Without
Mk. II	750/850	450/550 ⎬	accelerator
Mk. II*	750/850	450/550 ⎭	
Mk. III	750/850		Fixed accelerator.
Mk. III*	–	600	No accelerator.

TO LOAD.

(a) Grasp the pistol grip with the right hand with fingers clear of the trigger and pull back the trigger guard until the cocking catch lever can be pressed down with the thumb. Slide the trigger guard forward as far as it will go and then pull it back until retained by the cocking catch. The gun is now cocked.

(b) Feed in the belt – pass the tag of the belt through the feed block from the right and pull to the left as far as it will go. The gun is now ready to fire. Tuck the end of the tag into the metal chute.

TO FIRE.

Squeeze the trigger of the remote control firing gear. Firing will continue until the trigger is released or the end of the belt is reached. If the belt is expended, the gun must be cocked again before leading in a fresh belt.

TO UNLOAD.

With the gun cocked hold back the trigger guard, pull out the cover locking pin, raise the cover and hold open by the ring suspended near the gun; remove the belt, see that the chamber is clear, lower the cover and engage the locking pin. Pull back the trigger guard, slightly depress the cocking catch lever and ease the working parts forward under control. With the trigger squeezed pull the trigger guard back, release the trigger and draw the trigger guard right back until retained by the cocking catch.

Fig. 78.—See guns are unloaded.

PRECAUTIONS.

(a) Always treat the gun as loaded until proved otherwise. Cock the gun, open the cover, and see that the chamber is clear.
(b) Do not fire the working parts forward when the gun is unloaded, unless absolutely necessary. Ease them forward (see "To Unload").
(c) The gas cylinder is very easily damaged. Avoid the following:—
   (i) Attempting to remove or replace the barrel when the gun is not cocked.
   (ii) When replacing the barrel, knocking the cylinder against the body of the gun.
   (iii) Firing the working parts forward with the barrel retainer disengaged.
(d) Although the parts of the gun are designed to be interchangeable, experience has shown that the components of each particular gun should be kept together and not assembled to any other gun, except in an emergency.

STRIPPING (see Fig. 79).

Do not strip the gun further than necessary. For cleaning, maintenance and examination the following procedure is adopted:—

(a) Lift off the rear baffle plate.
(b) Raise the carrying handle (46) until just clear of the lug on the right side of the body and push the barrel retainer (44) forward until clear of the slides in the body (2). Raise the carrying handle to the vertical position, lift the rear of the barrel (41) and push it forward until the guides on the barrel sleeve (40) are disengaged from, the guides at the front of the body.
(c) Pull out the cover locking pin as far as it will go. Remove the cover (85).
(d) Press in the belt guide catch (if fitted) and lift the belt guide from its housing in the body.
(e) Lift the feed block from the body and slip out the feed slide.
(f) Remove the breech block (31) by lifting the rear and sliding it out backwards.
(g) Remove the accelerator (1) (if fitted) by pulling out the accelerator arm plunger from the body, turning it to the vertical position (upward for Mk. I guns and downward for Mk. II and II* guns) and lifting the accelerator from its seating in the body.
(h) Ease the working parts forward (see "To Unload").
(i) Press the return spring guide block (34) forward (1 in. approx.) and with a lift, remove the return spring guide and return spring (33) from the body.
(j) With one hand on the piston extension and the other at the rear end of the barrel extension (27), lift out the piston (32) and barrel extension. Slide the piston out of the barrel extension.
(k) Lift out the feed lever.
(l) Raise the trigger guard catch. Grasp the pistol grip, squeeze the trigger (56), draw the trigger guard to the rear as far as it will go, release the trigger and again draw the trigger guard to the rear until it comes away from the gun.
(m) Return the barrel. With a punch or the point of a bullet, depress the gas cylinder sleeve spring (50) and, using the spanner end of the combination tool or an adjustable spanner, rotate the gas cylinder sleeve (51) until it is free of

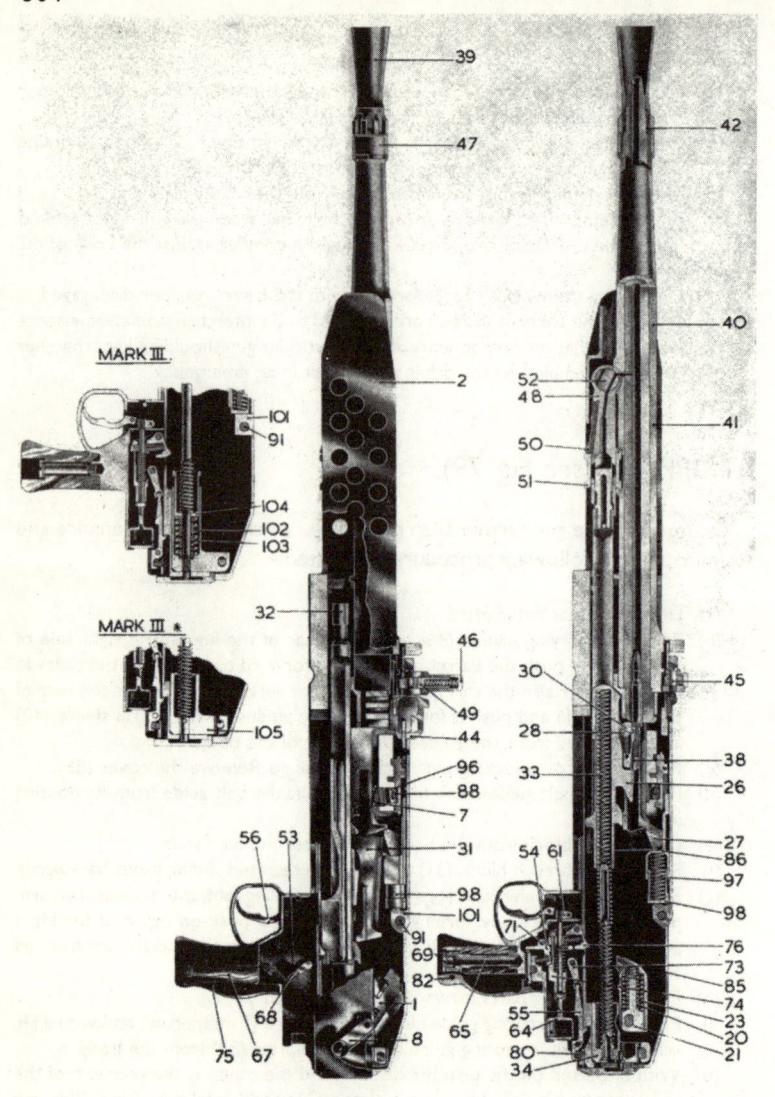

Fig. 79.—7·92 mm. Besa machine-gun—sectioned.

its housing in the barrel. Swing the rear end of the, gas cylinder (48) away from the barrel sleeve (40) until the gas cylinder becomes detached from the barrel sleeve. Tap the gas regulator (52) out of the cylinder with a copper hammer or brass drift. Slip off the gas cylinder sleeve (51) and spring.

(n) Turn the breech block (31) upside down, lift the front end of the extractor stay (28) until it is disengaged from the extractor (30) and remove it, together with its spring. Lift out the extractor.

(o) Finally, turn the breech block upright, press the firing pin retainer downward with a punch or the point of a bullet. The spring will force the firing pin (38) out.

ASSEMBLING.

(a) Assemble the spring to the firing pin (38), and with the retainer still down, press the pin and spring into the breech block (31), taking care that the slot in the pin is facing the retainer. Push home the retainer.

(b) Turn the breech block upside down and slide the extractor (30) into its guides. Assemble the spring to the extractor stay (28) and place them in the breech block, rear end first, with the projection uppermost. Press in the front end of the stay until it is retained.

(c) Replace the gas cylinder sleeve spring in the gas cylinder (48). Slip on the gas cylinder sleeve (51), with the interruptions on the opposite side to the spring. Depress the spring and push down the sleeve so that it holds the spring depressed. Insert the gas regulator (52) into the cylinder. Engage the flange on the gas cylinder in its housing in the barrel sleeve (40) and swing the cylinder to the rear until it lies along the barrel. With the combination tool or a spanner, rotate the gas cylinder sleeve until it engages in its housing in the barrel sleeve.

(d) Raise the trigger guard catch and engage the flanges of the trigger guard with the guides in the body; with the cocking catch thumb-piece depressed, slide forward the trigger guard, keeping the fingers clear of the trigger (56). Release the cocking catch thumb-piece and drop the trigger guard catch. Pull the trigger guard back until the cocking catch (55) engages with the body.

(e) Replace the feed lever and swing its upper arm out to the right.

(f) Slide the upper flanges of the piston extension (26) into the lower groove of the barrel extension (27). With the piston in the forward position, lower the piston and barrel extension into the body.

(g) Assemble the return spring (33) over the return spring guide (34). Grasp the top of the return spring guide block with the right hand and insert the free end of the spring into the piston extension (26). With the left hand supporting the spring, force the return spring guide forward until the guide rod enters the body and then press the guide downwards and release. See that it is correctly positioned.

(h) Cock the gun, keeping one hand pressed down on the barrel extension.

(i) Replace the accelerator (1) (if fitted) and engage the plunger.

(j) Replace the breech block (31), making sure that it is properly settled down on to the piston extension (26).

(k) Slip the feed slide into the feed block and position the left edge of the slide itself in line with the left edge of the feed block. Lower the feed block into the body and ensure that the stud on the slide is engaged in the slot of the feed arm.

(l) Replace the belt guide in the body and press it downwards until the catch (if fitted) engages.

(m) Engage the cover (85) with the trunnions on the body, close it and push in the cover locking pin.
(n) Take hold of the barrel carrying handle (46) and raise the rear end of the barrel. Keeping the gas cylinder (48) clear of the body (2), engage the guides on the barrel sleeve (40) with the guides at the front of the body, pull the barrel to the rear and lower the breech end into the barrel extension (27). Push the carrying handle over to the right so that it rests on the ramp; knock back the handle with the hand and push it down into the locked position.
(o) Replace the rear baffle plate.
(p) Ease the working parts forward (see "To Unload").
(q) Test the gun for correct assembly by cocking and easing the working parts forward again.

STRIPPING IN ACTION.

When the gun is mounted, the barrel, piston, barrel extension, feed lever and gas cylinder cannot be removed.

The following parts can be removed if replacement or repair is required in action:—

In each case the cover must be opened first.

Breech Block and Components – With the gun cocked. See "Stripping" (f), (n) and (o).

Return Spring and Guide – With the working parts eased forward, after removal of the accelerator (if fitted). See "Stripping" (g) and (i).

Belt Guide – By pressing in the catch (if fitted) and lifting out.

Feed Block and Feed Slide – By lifting out, after removal of the belt guide.

Accelerator (if fitted). See "Stripping" (g).

Trigger Guard – See "Stripping" (l).

Feed and Retaining Pawl Springs – These can be replaced in emergency by manipulation, using a small screwdriver.

STOPPAGES.

Breakages are rare with the Besa machine-gun. Correct handling, attention to maintenance and periodical examination are essential to ensure freedom from stoppages, most of which are due to faulty handling, careless preparation or lack of inspection.

It is essential to protect the gun and ammunition from rain, dust, and extreme cold, and efforts should be made to avoid overheating.

It is advisable to "run-in" new guns during training to bring to light any defects which can be remedied before battle.

Parts of a gun found reliable should not be exchanged with other guns, except in an emergency.

Stoppages fall within the following groups, and will be dealt with in subsequent sections:–

(a) Misfires and related stoppages.
(b) Stoppages due to the cartridge not being driven clear of the belt.
(c) Extraction and ejection stoppages.
(d) Feed stoppages.

Each stoppage has "indications" by which it may be diagnosed. Many stoppages give the same indication and must be distinguished by elimination. Many stoppages have associated stoppages which give an indication as to the cause. "Immediate Action" (I.A.) is the action carried out by the gunner to make the gun fire again in the least possible time, and should be almost instinctive.

PRECAUTIONS WHEN CLEARING STOPPAGES.

Serious damage to guns and injuries to gunners will result upon careless handling when loading or clearing stoppages, and the following precautions should be observed:–

(a) Keep the fingers clear of the trigger when cocking.
(b) Always cock the gun or hold back the working parts by the trigger guard; open the cover and do not release the trigger guard until engaged by the cocking catch.
(c) When cocking, once the action of drawing back the working parts has been commenced, they must not be allowed to slip forward even if they cannot be drawn right back. Carelessness in this action will result in a double feed, with a possibility of a round in the breech being fired by a round in the belt when the cover is raised, and usually results in a bullet lodged in the bore.
(d) If a lodged bullet is suspected the bore must be cleared before firing to prevent the barrel being bulged and the breech block fractured.
(e) Do not support the cover with the head, but suspend the cover from the roof of the vehicle by the ring provided.
(f) Clear the chamber as soon as possible, as an overheated gun will give a "cook-off" with the same results as in (c).
(g) Never fire the working parts forward even if the gun is clear unless the cover is closed and locked. (A)
(h) When removing the belt, ensure that the exit guide – if not fitted with a retaining catch – is not drawn out of place.
(i) When replacing the belt in the gun see that the first round is in line with the chamber.

Any unusual or persistent stoppages should be reported and the gun handed in to an armourer, together with defective parts and samples of fired and unfired ammunition and belts. A statement should be made of the circumstances, and particulars on the ammunition boxes should be quoted.

THE 2-IN. BOMBTHROWER, MARK IA*
(See Figs. 80 and 81)

A two-inch bombthrower is fitted to fire smoke-emitting bombs from inside the A.F.V. Alignment for aiming is obtained by swinging the turret on to the target, using the turret sighting vane as a sight. Range for aiming is obtained by adjusting a gas regulator valve fixed to the side of the bombthrower. Maximum range, 150 yards.

There are two types of bomb, the type being clearly marked on the body. The bomb smokes an area of about 33/40 sq. yds. in still air.

REMEMBER:–

(1) To aim UP WIND and let the smoke drift to the target.
(2) That a HIGH WIND will move and DISPERSE smoke quickly.
(3) To RELOAD.
(4) The bombthrower must always be in the CLOSED POSITION, except when actually loading or unloading.

LUBRICATION AND MAINTENANCE.

All working parts should be kept thoroughly cleaned and oiled and free from burrs. Before new lubricants are applied, the old should be removed to prevent grit being retained. After lubrication, *operate the mechanism*.

Lubricants:–

(1) Mineral jelly – Generally.
(2) C.70 (or M.160) oil – All bright parts.
(3) Grade 2 kerosene oil – For removing the above.

No parts will be burnished.

Keep the bombthrower free from rust and the bores slightly oiled.

Remove fittings frequently to check operation, but don't use unnecessary force.

Only ARTIFICERS should remove BURRS.

Report any flaw or crack at once.

Before firing – clean and dry the bore.

After firing – wash the bore in hot water (if available), and oil when cool.

Paraffin can be used on very dirty bores, and if no water is available.

Do NOT use SODA in any form to clean the bore.

All spare parts should be tested for interchangeability as soon as possible after receipt.

AMMUNITION

MARKINGS ON Q.F. 6-PR. 7-CWT. AMMUNITION.

(a) **General.**
- (i) Ammunition issued to the Service is marked to facilitate identification and to ensure segregation in store and transport. Markings also ensure that the correct types are used and assist in tracing defects in design and manufacture.
- (ii) Care should be taken in handling ammunition, to avoid damage to the markings.
- (iii) Q.F. fixed ammunition is batched for the purpose of recording the components used in the make-up of the cartridge, and ammunition of the same batch should give consistent shooting. Each batch contains a propellant charge of one lot only, i.e. it was all made at the same place and time, but the fuses may be of more than one lot.
- (iv) Batches are distinguished by consecutive numbers, preceded by the appropriate letter, e.g. "Batch E.I" denotes the first batch of shell ammunition. When it is found necessary to use more than one lot of fuses in a batch, it is divided into sub-patches, as follows:-"Batch E.I" – containing rounds with first lot of fuses; "Batch E.I-A" – containing rounds with second lot of fuses; "Batch E.I-B" – containing rounds with third lot of fuses.
- (v) Batches will be stored separately and will be so arranged as to avoid dividing a batch or sub-batch.
- (vi) A label is affixed to the inside of each box giving particulars of the components contained in the ammunition. When it is necessary to replace original

Fig. 87.—Don't be careless with ammunition.

Cartridge, Q.F., 6-pr., 7 cwt.	Case empty.	Primer, percussion.	Charge.	Tinfoil.	Tracer.	Projectile.	Fuse.
Armour-piercing shot, Mk. IT, foil	Mk. I	No. 15, Mk. II	1 lb. 13 oz. 8 dr. W.T. or W.M.T. Cordite	1 dr.	Internal	Shot, A.P., Mk. IT, IIT, IIIT or IVT	—
Armour-piercing shot, Mk. IIT, foil	,,	,,	2 lb. 5 oz. 6 dr. N.H. 033	,,	,,	Shot, A.P., Mk. IT, IIT, IIIT or IVT	—
Armour-piercing shot, Mk. IIIT, foil	,,	,,	1 lb. 13 oz. 8 dr. W.T. or W.M.T. Cordite	,,	,,	Shot, A.P., Mk. VT, VIT or VIIT	—
Armour-piercing shot, Mk. IVT, foil	,,	,,	2 lb. 5 oz. 6 dr. N.H. 033	,,	,,	Shot, A.P., Mk. VT, VIT or VIIT	—
H.V. armour-piercing shot, Mk. IT, foil	,,	,,	2 lb. 7 oz. N.H. 033	,,	,,	Shot, A.P., Mk. IT, IIT, IIIT or IVT	—
H.V. armour-piercing shot, Mk. IIT, foil	,,	,,	2 lb. 7 oz. N.H. 033	,,	,,	Shot, A.P., Mk. VT, VIT or VIIT	—
Armour-piercing shot, with cap, Mk. IT, foil	,,	,,	1 lb. 13 oz. 8 dr. W.T. or W.M.T. Cordite	,,	,,	Shot, A.P., with cap, Mk. VIIIT	—
Armour-piercing shot, with cap, Mk. IIT, foil	,,	,,	2 lb. 5 oz. 6 dr. N.H. 033	,,	,,	Shot, A.P., with cap, Mk. VIIIT	—
Practice, Mk. IT, foil	,,	,,	2 lb. 5 oz. 6 dr. N.H. 033	,,	,,	Shot, practice, Mk. IT or IIT	—
Practice, Mk. IIT, foil	,,	,,	1 lb. 13 oz. 8 dr. W.T. or W.M.T. Cordite	,,	,,	Shot, practice, Mk. IT or IIT	—
Practice, Mk. IIIT, foil	,,	,,	2 lb. 5 oz. 6 dr. N.H. 033	,,	,,	Shot, practice, Mk. IIIT, IVT, VT or VIT	—
Practice, Mk. IVT, foil	,,	,,	1 lb. 13 oz. 8 dr. W.T. or W.M.T. Cordite	,,	,,	Shot, practice, Mk. IIIT, IVT or VT	—
Practice shot, flat-headed, Mk. IT, foil	,,	,,	Reduced charge, 1 lb. 1 oz. 6 dr. W.M. 061 Cordite	,,	,,	Shot, practice, flat-headed, Mk. IT, IIT, IIIT or IVT	—
Practice shot, flat-headed, Mk. IIT, foil	,,	,,	Reduced charge, 1 lb. 5 oz. N.H. 025	,,	,,	Shot, practice, flat-headed, Mk. IT, IIT, IIIT or IVT	—
H.E. shell, Mk. I, foil	Mk. IM	,,	2 lb. 5 oz. 6 dr. N.H. 033	,,	,,	H.E., 6-pr., Mk. VII	Percussion, D.A. No. 244
H.E. shell, Mk. IIT, foil	,,	,,	2 lb. 5 oz. 6 dr. N.H. 033	,,	Shell No. 13	H.E., 6-pr., Mk. XT	Percussion, D.A. No. 244
Blank	,,	No. 20	15 oz. blank L.G., G.12 or R.F.G.2	—	—	—	—
A.P.C., P.C.	—	—	—	—	Internal	Shot, A.P.C., B.C., Mk. XT	—

components by those of other lot numbers, the letter "X" will be appended to the batch or sub-batch numbers on the box. It will also be stencilled on the side of the cartridge case when the fuse is changed. The letter "X" denotes that the box contains components other than those originally packed.

(vii) If possible, ammunition will be repacked in the boxes from which it was removed. Failing this, the batch or sub-batch number on the box must be amended.

7.92 MM. BESA AMMUNITION.

The 7.92 mm. M.G. takes a rimless cartridge comprising case, cap, charge and bullet.

The base of the cartridge is stamped with the mark, contractor's initials or recognised trade mark, and the last two figures of the year of manufacture. The annulus is coloured to indicate the character of the cartridge.

The following marks of cartridges are issued:—

Cartridge S.A. Besa 7.92 mm. Mk. I.Z and Mk. II.Z annulus coloured purple.
Cartridge S.A. armour-piercing 7.92 mm. Mk. I.Z and Mk. II.Z annulus coloured green. Cartridge S.A. tracer 7.92 mm. Mk. I.Z and Mk. II.Z annulus coloured red.
Cartridge S.A. incendiary 7.92 mm. Mk. I annulus coloured blue.
Blank drill and dummy cartridges are available.

The cartridges are carried in belts holding 225 rounds, the filled belts being stored in wooden boxes holding two tinned-plate boxes or liners from which the gun is fed direct. Each liner holds two belts which may be loaded, with ball, ball and tracer, A.P., or incendiary or a combination in definite proportions.

The boxes and liners are labelled to state the contents.

2-IN. BOMBTHROWER AMMUNITION.

The ammunition for the 2-in. bombthrower consists of two types of cartridge:—

Cartridge, 2-in, Bombthrower (18 grain ballistite), Mk. I;
Cartridge, 2-in. Bombthrower (42 grain G.20);

and two types of smoke bomb:—

Bomb, Smoke, 2-in. Bombthrower, Mk. I and II;
Bomb, Smoke, Bursting, 2-in. Bombthrower, Mk. III.

Both cartridges and bombs have the usual markings, as have also the boxes and packages. The boxes hold eighteen rounds in packages of six and are painted green to denote smoke ammunition. The smoke bombs are also painted green, a red band denoting that they are filled and a white band if they contain white phosphorus.

The Mk. II smoke bomb differs from the Mk. I in that a short delay occurs in the emission of smoke. The Mk. III contains as a smoke composition white phosphorus and is assembled with a D.A. percussion fuse covered with a safety cap, which must be removed before firing.

CARE AND PRESERVATION OF AMMUNITION.

Ammunition is not foolproof — be careful of it. Keep it away from water, rain, damp, direct sunshine and gas, whether it is stored in the vehicle or in packages. Try to avoid extreme heat or cold. Ordinary temperatures will not affect ammunition.

Load or unload ammunition in dry weather, or under cover. Loads in open vehicles should be covered by tarpaulin. See that the "B" echelon ammunition is properly sheeted. If the ammunition package has been exposed to wet in transit, make sure there is no water inside. If so, dry the contents of the package and carefully repack.

Don't throw packages about.

Always stack ammunition under cover, and provide for a free circulation of air. If possible stack ammunition packages on battens. Keep them 12 in. clear of walls, and leave passages of at least 18 in. between stacks. Maximum height of stacks allowed is 12 ft. Try to keep one end of each package visible.

On receipt of ammunition, examine all labels carefully to see that the ammunition is suitable and sort it into kinds if there are several in the consignment. Look for broken seals, open any boxes which appear to have been tampered with, and ensure that the contents correspond with the labels and are still serviceable.

Ready-use ammunition – The amount of ammunition unpacked from boxes must be governed by the stowage room in the vehicle. Ammunition must not be unpacked unless necessary, and the opening of sealed liners containing M.G. ammunition should be deferred until the last possible moment, as the liners cannot be resealed.

Ammunition stowed in a vehicle must be inspected weekly and before action, by the crew commander. If possible, ammunition ready for use, should be "turned over" by expenditure. Do not oil ammunition as this leads to the accumulation of dust.

The clip, No. 36 Q.F., protecting the primer in the cartridge case, *must be kept in position* until the round is required.

Inspection of ammunition – Every round must be inspected before using. When time permits, rounds should be gauged by being offered into the breech, but *the striker case must first be removed* from the breech block.

The *projectile* must be clean and free from dust or grit. It must be secured tightly in the cartridge case and the driving band must be undamaged. If H.E. shell, the fuse must be tight and no signs of exudation of the shell filling be visible.

The *cartridge case* must be clean, have no cracks or serious dents. Short cracks at the mouth up to ¼ in. may be disregarded, cracks elsewhere necessitate rejection. The rim must be undamaged, and the primer must be flush with the base and screwed tightly home; if sunken or projecting it must be rejected.

Damaged ammunition – Damaged or faulty ammunition should be returned immediately in its original boxes, care being taken to see that the labels are not missing, as these are the means of tracing faulty batches to their source. If ammunition has to be returned in boxes other than the originals, the labels must be altered to correspond with the contents. Reports of faulty ammunition must always be accompanied by an extract from the box labels giving the batch dates of the cartridges.

Ammunition salvage – Fired cases, spent belts, felt strips and packing pieces, empty liners and boxes are all required as salvage and must be returned whenever circumstances permit. Live rounds must be separated from empty cases. If conditions do not permit sorting, the boxes of empties should be conspicuously labelled to indicate that live rounds are present.

Don't – Hammer or tap packing cases containing ammunition.
        Tamper with a round.
        Use live rounds for drill purposes.
        Have rusty rounds.

SIGHTING TELESCOPES

TESTING AND ADJUSTING.

Upon the correct adjustment of sights depends the effectiveness of fire and the frequent testing of sights is of great importance.

Sights should be tested after removal and re-insertion in the mounting before firing and especially after firing or travelling, when shocks and vibration may have disturbed the telescope.

(a) To Test 6-Pr. Gun.

Select a well-defined object, approximately 1,000 yards distant, on which to lay the gun. Fix fine wires horizontally and vertically across die front face of the muzzle of the 6-pr. gun in the four axis lines marked. It will be necessary to remove the muzzle weight if fitted. Remove the striker case. Align the bore of the gun on the distant object, using the intersection of the cross-wires as a foresight and the hole in the firing hole bush as a backsight and lock the mounting in position. To obtain accurate alignment the eye should be drawn back as far as possible from the firing pin hole without losing sight of the aiming mark. Re-check the alignment several times and ensure that the gun has not moved. If the sights are correct, the intersection of the vertical cross-line in the telescope and the zero line should be on the same spot of the distant object as the gun. The short horizontal line above the intersection of the cross-lines indicates zero. If there is an error, adjustment will be necessary.

(b) M.G.

Remove the breech block and accelerator (if fitted) from the M.G. (see "Stripping") so as to give a view through the bore of the barrel and the sight hole at the rear of the body. If the mounting is correct the gun should lie on the same point of aim as the 6-pr. gun. If deviation is present it should be reported to a gun fitter immediately.

(c) To Adjust the Telescope.

Slacken the four adjusting screws. The diaphragm on which the range scales are marked can be moved vertically by pressing the top or bottom adjusting screw, and horizontally by pressing the right or left screw. Tighten the screws when adjustment is correct.

(d) Precautions.

It is essential that the sights be taken exactly in the centre of the bore of the gun. Considerable error will arise if the eye is not in the line of the axis.

Movement of the gun after alignment and prior to the completion of the test will negative the result. Such movement may be produced by movement of the tank as a whole, jolting of the gun or rocking or rotation of the turret in any way. The gun and telescope must be aligned by the same man and the tests must be rechecked.

(e) Testing the Parallax.

Telescopes should be tested periodically for parallax. Align the telescope on a well-defined object approximately 500 yards distant. Move the head vertically and horizontally while looking through the telescope.

If the telescope is correctly adjusted, there will be no apparent movement of the cross-wires. If the cross-wires appear to move with the head, adjustment is needed and the telescope will be sent for repair.

No. 20 MOUNTING.

The method of testing is similar to that described for the co-axial mounting and consists of laying the M.G. on a distant object. The line of sight through the intersection of the vertical cross-line in the telescope and the zero line should be on the same spot of the distant object as the gun. If not, the telescope must be collimated by shortening or lengthening the compression link below the M.G. cradle for obtaining lateral adjustment and by tightening or slackening the adjusting screw at the left of the telescope housing for obtaining vertical adjustment. After adjustment the lock nuts of the compression link and of the adjusting screw must be screwed home.

The No. 35 telescope fitted to the No. 20 mounting is removable, for cleaning or renewing the head, by operating the locking bolt to release the telescope clamp. The telescope can then be withdrawn downwards from its mounting.

MACHINE GUN, NO. 20, MOUNTING

The M.G. mounting is of the gimbal type and is situated forward in the vehicle, on the left of the driver. The inner mantlet and cradle in which the M.G. is mounted moves on trunnions both vertically or horizontally in an outer mantlet. The vertical trunnions give a traverse, left or right, of 22½ degrees, and the horizontal ones and elevation or depression of 12½ degrees. The No. 35 Mk. I sighting telescope is fitted on the left of the mounting and is moved by parallel link and lever motion actuated from the gimbal ring and cradle, to correspond with the direction of fire. A split spherical joint, held in a bracket bolted to the hull, when tightened by a handwheel, clamps a slide bar to lock the mounting in position. Aim is controlled by pressure on a handle attached to the underside of the cradle and the firing grip of the M.G. The M.G. is fired by finger pressure on the trigger.

The sighting telescope is removable from its mounting by releasing the locking bolt immediately above the eyepiece.

MAINTENANCE.

The working surfaces and mechanism must be kept free from dust and well lubricated, particular lid to the bearing surface of the link and lever mechanism by means of which motion of the M.G. is imparted to the sighting telescope. Special attention must be given to the table at the left side of the telescope mounting, as dust or dirt thereon may cause the slipper to give an incorrect movement to the table and affect the alignment of the sighting telescope.

Lost motion at any of the points of connection of the telescope link and lever mechanism must be eliminated, as slackness at any joint will seriously affect the correct movement of the telescope in synchronism with the M.G.

Also available from Amberley Publishing

SPITFIRE MANUAL 1940

AIR MINISTRY

'As a reminder of times as a fighter pilot in WWII this book is a must' *AIRCREW ASSOCIATION*

Available from all good bookshops or to order direct
Please call **01453-847-800**
www.amberley-books.com

Also available from Amberley Publishing

AIR MINISTRY

LANCASTER MANUAL 1943

Available from all good bookshops or to order direct
Please call **01453-847-800**
www.amberley-books.com